Year 4/Primary P5

CW01651231

ncluding Lower-achievers in the Maths Lesson

Sue Atkinson

HOPSCOTCH EDUCATIONAL PUBLISHING

Contents

Published by
Hopscotch Educational Publishing Ltd,
29 Waterloo Place,
Leamington Spa CV32 5LA
Tel: 01926 744227

© 2000 Hopscotch Educational Publishing

Written by Sue Atkinson
Series design by Blade Communications
Illustrated by Bernard Connors
Printed by Clintplan, Southam

ISBN 1-902239-58-X

Sue Atkinson hereby asserts her moral right to be identified as the author of this work in accordance with the Copyright, Designs and Patents Act, 1988.

All rights reserved.
This book is sold subject to the condition that it shall not, by way of trade or otherwise, be lent, hired out or otherwise circulated without the publisher's prior consent in any form or binding or cover other than that in which it is published and without a similar condition, including this condition, being imposed upon the subsequent purchaser.

No part of this publication may be reproduced, stored in a retrieval system, or transmitted, in any form or by any means, electronic, mechanical, photocopying, recording or otherwise, without the prior permission of the publisher, except where photocopying for educational purposes within the school or other educational establishment that has purchased this book is expressly permitted in the text.

The author would like to thank the many teachers and children from the following schools who trialled these activities:

Upton House School, Windsor

Benyon Primary School, South Ockenden, Essex

St. Josephs RC Primary School, Stanford le Hope, Essex

West Thurrock Primary School, Thurrock, Essex

Christ Church C of E Primary School, South Croydon

Haymerle Special School, London

Bishop Perrin School, Richmond

Sheen Mount Primary School, Richmond

Summerbee Junior School, Bournemouth

St Ives First School, St Ives, Ringwood, Hampshire

Mudeford Junior School, Christchurch, Hampshire

Linwood School (Special), Bournemouth

In the section on Inclusion, The National Curriculum states that schools have a responsibility to provide a broad and balanced curriculum for all pupils. It sets out three principles that are essential to developing a more inclusive curriculum:

A: Setting suitable learning challenges

B: Responding to pupils' diverse learning needs

C: Overcoming potential barriers to learning and assessment...

Including Lower-achievers in the Maths Lesson is a series of books that gives support for including in the daily maths lesson those children in the class who, for whatever reason, are struggling and therefore often failing.

The books can be used alongside the ***Developing Numeracy Skills*** books also published by Hopscotch Educational Publishing Ltd. You will find a cross-reference to these books on the chart on page 5.

CHAPTER CONTENT

The overall learning objectives

Each chapter has overall learning objectives that are based on the key objectives in the *National Numeracy Strategy Framework for Teaching.*

The grid on page 5 shows where these key objectives are covered within the chapters in this book. The lessons in this Year 4/P5 book are based on the Number Key Objectives for Years 2–4 (P3–5) to be found in the *Framework for Teaching.*

The assessment focus

Each chapter is divided into 'sessions' – a set of activities that might last for one or more days. The assessment focus for each session is based on the key objectives and broken down into competencies that the children are expected to achieve within those key objectives.

An individual assessment chart can be found on pages 119 and 120.

With the whole class

This section includes activities that are suitable for everyone in the class but they focus on the work for the lower-achievers that follows. Therefore they use easy numbers so that the lower-achievers can cope with the maths concept being taught.

You can repeat the whole-class starter(s) over several days or choose another starter that focuses on a similar learning objective.

Some suggested vocabulary is given, but you should teach with your *National Numeracy Strategy Mathematical Vocabulary* to hand.

With the lower-achievers

These are activities that follow on from the whole-class starter and you can use them:

- in the middle of the lesson for children who are under-achieving
- during the next week or two to consolidate the concepts
- when you come back to that topic next term.

Within this section is a selection of activities, some to be done with adult guidance and some which can be done by selected children independently.

With adult support

Many of the activities require equipment. We use equipment so that we are developing the children's mental images and things such as number lines, 100 squares and cubes are absolutely crucial. Many of the activities are games as these can motivate and keep the children involved. You will need to train your children to play games if they are not used to them.

Once taught, some of these activities can be teacher-independent. Often there are examples of questions and other things to say to the children so that their vocabulary is extended.

In addition to the adult support during the activities, try to plan for times when an adult can give a few minutes 'catch up' time to a group of children at odd moments during the day, for example five minutes at the end of assembly or at the start or end of the day. (If possible, take children from more than one class for 'catch up' times to maximise the time of an adult helper.)

Teacher-independent activities

These activities are suitable for lower-achievers to do on their own, depending how well they work independently. Again, many of them will require the use of maths equipment.

Plenary session

In this section a few questions are included to help lower-achievers reflect on their work, and sometimes there is an additional activity. Many of the whole-class starters can also be used as activities here.

Keep plenary sessions varied and interesting. They are times to sort out misconceptions, and times to generalise about what has been done. They are not just 'show and tell' times!

The photocopiable resource sheets

Some of the activities suggested in the teachers' notes require the use of resource sheets. A minimum of words is used on these, so helpers will need to have instructions on what to ask the children to do on them. Many of the sheets are blanks, requiring the teacher or adult to write in tasks at a suitable level of ability.

On the more complex sheets for younger children there is a space at the bottom that contains notes for the adult. Sometimes this space is blank. This has been done deliberately as the teacher or adult may wish to add a few notes for another adult to use with a specific child or group. Likewise the sheets may be sent home to the parents/carers and this space can be used for instructions to them.

Some of the artwork on the resource sheets is on the same theme as the ***Developing Numeracy Skills*** books. However, some of it is more varied to provide interest for the children. We have also provided 'clip art' sheets at the end of the resource sheet section, so that you can cut and stick pictures onto the children's sheets in order to change the appearance of the sheets when you want to use them more than once. This gives the sheets new life and gives the children plenty of practice to consolidate.

A general point about using the sheets with lower-achievers

Although the resource sheets provided in this book can be used several times, remember that it isn't recording maths that is important; it is whether the child can understand the concept. So use the sheets only when they are needed, such as for a teacher-independent activity or for assessment.

Spinners

Some of the games in the book need a spinner that works by trapping a paper-clip in the middle of the spinner then flicking the paper-clip round with a finger. Spinners are quieter than dice and give more choice.

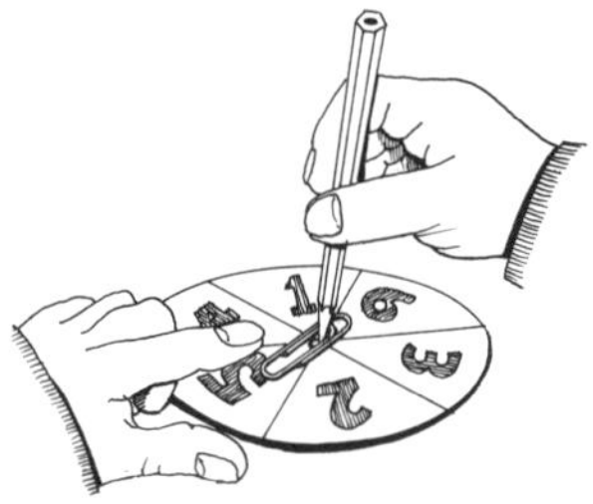

Number lines

Have a large wall number line and 100 square up in each class. Remember that 100 squares are suitable for looking at patterns, such as for 10s, but are not ideal to do calculations on because children tend to get lost as they move from the end of one line to the start of the next. Number lines are much easier to use for calculations.

And finally, remember

We want children to feel positive and to feel that they are achieving well in maths. Working endlessly on activity sheets can be boring, and could even make a child feel they are failing. Use games where possible to reinforce concepts and skills. Be generous with praise.

This grid is made up of the key objectives for Years 2, 3 and 4 plus other crucial learning, and shows the chapter/s where these objectives are covered in this book. The grid also shows where the same area of learning can be found in the ***Developing Numeracy Skills*** series.

KEY OBJECTIVES	CHAPTERS	DNS – BOOK/CHAPTER
Read, write and order whole numbers to 1000 and know the value of each digit	1, 6, 7	Year 4 – Chapters 1 – 3, 5, 9 – 12
Count forward or back in 10's or 100's from any two- or three-digit number	1	Year 4 – Chapters 1 – 3
Use symbols correctly: '=', '<' and '>'	1	Year 4 – Chapter 2
Round to the nearest 10 or 100	1	Year 4 – Chapters 1 – 3
Recognise and extend number sequences	2	Year 4 – Chapters 3, 7, 11, 14
Recognise units and other simple fractions and mixed numbers	6	Year 4 – Chapters 9 – 12
Order simple fractions and be able to find fractions of numbers	6	Year 4 – Chapters 9 – 12
Understand decimal notation (tenths and hundredths) in the context of money	6	Year 4 – Chapters 9 – 12
Recognise equivalence of simple fractions	6	Year 4 – Chapters 9 – 12
Know by heart all addition and subtraction facts for each number to 20	1, 3	Year 4 – Chapters 1 – 6
Have a range of calculating strategies	3	Year 4 – Chapters 4 – 6
Use known number facts and place value to add or subtract mentally, including pairs of two-digit numbers	1, 3, 7	Year 4 – Chapters 1 – 6, 11
Recognise that addition is the inverse of subtraction	3, 4	Year 4 – Chapters 4 – 6
Add/subtract mentally and on a number line a near multiple of 10	3	Year 4 – Chapters 4 – 6
Develop and refine written methods for addition and subtraction using a number line and easy column addition	3, 4	Year 4 – Chapters 4 – 6
Understand division as both grouping and sharing	5	Year 4 – Chapters 7, 8
Recognise that division is the inverse of multiplication	5	Year 4 – Chapters 7, 8
Know by heart facts for the 2, 5 and 10 times tables and derive quickly simple division facts	2, 5	Year 4 – Chapters 3, 7, 8, 11, 14
Find remainders after division	5	Year 4 – Chapters 7, 8
Be able to solve problems by choosing the appropriate number operation	7	Year 4 – Chapters 5 and 11
Reason about numbers	2, 7	Year 4 – Chapters 3, 5, 7, 11, 14

Place value, counting, ordering and rounding

Overall learning objectives

- Read, write and order whole numbers to at least 1000 and know what each digit in a number represents. (For decimal numbers see Chapter 6.)
- Understand and use the vocabulary of comparing and ordering numbers.

Key words

more than	round
less than	roughly
estimate	

100 squares

Assessment focus

- Can the children order numbers to 100, recognise the pattern of the 10's and 1's, read number names?

Resources

- large 100 square
- Resource sheets 1, 2, 3 and 4
- scissors, small squares of paper, Blu-tack, sticky notes, metre rule
- number name flash cards (for example 'forty' and 'five')

With the whole class

- Count to 100 and back to zero.
- Display a large 100 square and explore the patterns, for example the diagonal pattern in 11's and counting down the vertical pattern of 10's. Cover some numbers with sticky notes and ask for the hidden numbers. Say *"How do you know that is 64?"* and *"Explain the pattern as we go down the chart like this."*
- When you repeat the activity use number name flash cards.

With the lower-achievers

With adult support

Choose from:

1 Help the children to make their own number square on a large sheet of paper, using a metre ruler to make the rows and columns. Ask *"How many rows do we need?"* and *"How many spaces altogether?"* Allocate to each child some numbers to write on small squares of paper while you finish the large outline. They then Blu-tack their numbers into position. Observe who can place a number easily. Who can name their numbers? Who can count in 10's?

2 With the completed 100 square, start counting in 1's and 10's. Make sure all the children can count down the chart in 10's (3, 13, 23, 33, 43...). Make sure they count the number at the end of a row then go to the start of the next row. (100 squares are not always good for addition and subtraction because children tend to get lost as they change rows. For this reason, use number lines not 100 squares for calculations unless you want to focus on counting in 10's.)

3 Let the children remove some of the Blu-tack numbers, muddle them up and ask someone else to replace them. Who needs more help with this? Match the numbers to the number name flash cards. Observe – who needs more counting experience?

4 Use Resource sheet 4 to point to an empty square. Ask *"What number goes here?"*

Teacher-independent activities

Choose from:

1 Using Resource sheet 1, prepare enough 100 square jigsaws for each child or pair, cutting along the thick lines. Those who cannot manage a 100 square can use the 1–50 rectangle.

2 Use Resource sheet 2 to fill in the numbers on the 1–50 rectangle or the 1-100 square.

3 Using the 100 square on Resource sheet 2, you or the children can make further jigsaws. If they are struggling, make very simple jigsaws cutting just the vertical or horizontal lines to emphasise the patterns of the 10's and 1's.

4 Use Resource sheet 3 by filling in a few numbers and let the children fill in the rest!

5 Give the children copies of the blank square on Resource sheet 4 for them to fill in all the numbers, either 0–99 or 1–100. Continue beyond 100 using another copy of the sheet.

6 Using Resource sheet 4, give the children more experience by repeating any of the activities 1 to 4 above with a 0–99 square.

Plenary session

- *"I'm thinking of a number and it is in the row with five 10's and the units digit is an odd number. What could my number be?"*
- *"Everyone shut their eyes while I cover a number. What is my number? How do you know?"*
- *"Let's count in 10's from 61 until we reach 261."*

Washing lines

Assessment focus

- Can the children order numbers? Do they know the value of each digit? Can they use the language of 'more than' and 'less than'?

Resources

- number cards and large teacher arrow cards
- washing line and pegs
- Resource sheet 5
- a stopwatch and calculators
- Blu-tack
- base 10 equipment, number line and 100 square

With the whole class

- Give out a selection of number cards and work together to order them on a washing line, for example ordering 12, 22, 32, 33, 101, 121, 112, 201, 301. Ask *"Which is more, 12 or 21? How do you know?"* and *"Who can tell me a number that could come between 101 and 121?"*
- Use large teacher arrow cards to demonstrate numbers, for example 643. Ask the children to name the numbers and say numbers that are less than and more than that number.

6	0	0		4	0		3	=	6	4	3

- (A very noisy game, best played outside.) Give out a number card to each child in the class (a complete or incomplete sequence) and ask them to order themselves in a long line. Next, set a stopwatch and give out different numbers for the children to order themselves into, this time trying to beat the previous time. For example, ask them to order themselves into multiples of 10, such as 30, 90, 200, 4000, and so on.

With the lower-achievers

With adult support

Choose from:

1 Leave the washing line up and give two minutes of daily practice at ordering just a few numbers. Use the language of 'greater than' and 'less than'.

2 Check that 'teen' numbers (11–19) can be recognised and read correctly alongside 31, 51 and so on and ordered appropriately.

3 Carry out more work just with the 10's and 1's arrow cards, linking these numbers to their position on the number line and 100 square and demonstrating the numbers with some kind of base 10 equipment.

4 Play 'Space invaders' on calculators. Tell the children a two- or three-digit number to enter, for example 365. They must get rid of the 3 completely in one subtraction (ie –300). Watch out for those who try to do just –3. They should then 'shoot down' the 6 by subtracting 60 and so on. Eventually you can extend this to 1000's.

Teacher-independent activities

Choose from:

1 Shuffle up some numbers and give them to the group to be pegged in the right order on the washing line. Give them other sets of numbers to follow up, such as a complete sequence of

numbers, for example 1–20 or 10–25, or all the answers in the 10 times table, or all the hundreds (100, 200 … 1000).

2 Put the group into pairs. Prepare some sets of cards, one for each pair, provide a stopwatch and let them race to put the numbers in order. They can Blu-tack numbers to a board or record their numbers on Resource sheet 5.

3 On Resource sheet 5, fill in some numbers that go up in 1's or 10's. The children have to fill in the missing numbers.

Plenary session

- *"Make a number with the arrow cards that has 200 in it."*
- *"Which is more, 31 or 13?"*
- *"Let's count in 100's from 201."*
- *"Let's count in 10's around the circle from 349."*

Empty box game

Assessment focus

- Can the children understand 'more than' and 'less than'?

Resources

- several large 1–6 dice
- Resource sheets 6, 7 and 8
- number cards 0–9 and 10–100
- cards with '<', '>' and '=' signs
- a large 1–100 square or number line on display
- paper-clips and pencils
- an abacus

With the whole class

- Draw on the board four empty boxes with a 'greater than' (or 'less than') sign between them.

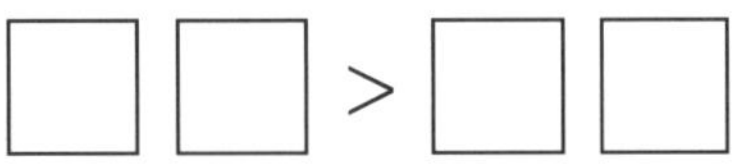

The purpose of the game is to make a true sentence, such as 61 > 34. Once a number is placed it cannot be moved. Throw the dice and ask the children where to position that digit in the empty boxes. If it is a 1 it might be wise to position it like this:

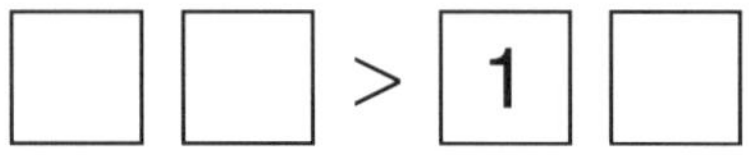

- Continue throwing the dice until all the empty boxes are full. Ask *"Why did you think it was best to put the 6 there?"* and *"How do we know that number is greater than this number?"*
- Repeat this but with six empty boxes to place 100's as well.

With the lower-achievers

With adult support

Choose from:

1 (Play this away from other classes if you are going to include the shouting out.) Using a selection of number cards from 10 to 100, give out a card to each child in the group. At a given signal they turn over their cards so that everyone can see them, and the first to shout out the highest number wins a point. The child whose card that is also wins a point. Order the cards and/or find the numbers on the number line. Collect in the cards and play again. Vary the game so that the winner has the lowest number.

2 Teach the game 'Three dicey numbers'. You need 2–4 players, a copy of Resource sheet 6 each and three dice.

Rules

i Take turns to throw all three dice and place them on the place value board at the top of the resource sheet so that you make the largest number possible. Say the number.

hundreds	tens	ones

"My number is five hundred and thirty two."

ii Put your number on the first row of arrow cards, carefully putting in the right number of zeros.

iii The next player throws the three dice.

iv When everyone has their first number, compare all the numbers. The player with the largest one scores 100 points.

v The winner has the highest score at the end.

Variation 1– Play so that the player with the smallest number wins.

Variation 2 – Play with some dice that have zeros on them so that you can make a number that is just tens and units.

Observe – Who names numbers confidently and uses 'more than' and 'less than' accurately?

3 Teach the game 'Abacus hundreds'. You need 2–4 players, Resource sheet 7, a paper-clip and a pencil to work the spinners (the spinners are at the bottom of the sheet) and an abacus (or use three different colours of Unifix).

Rules

i Take turns to spin your spinners, one at a time. Draw your 'beads' on your first abacus in the correct columns. Write your numbers under the columns and name your number.

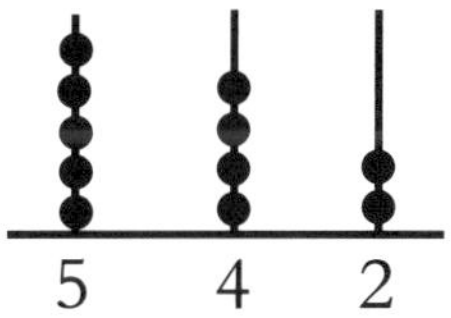

ii When each player has done their first abacus drawing, compare the numbers. The player with the highest number scores 100 points.

Teacher-independent activities

Choose from:

1 The children can play the dice game that was played in the whole-class starter but with four empty boxes only. They can work in pairs, each pair with a dice and Resource sheet 8 to record what they do. They tick either the 'yes' or the 'no' depending on whether they make a true sentence or not. (At the plenary session, order some of the numbers they have made on the washing line, noting that the greater numbers are positioned to the right of the smaller ones.)

2 There is space on Resource sheet 8 for the children, or you, to insert additional boxes so that the game can be played with 100's as well.

3 Vary the game a bit by giving pairs a set of 0–9 number cards face down between them. Each child selects two cards and makes the largest possible number. The player with the largest number wins a point.

| 6 | | 4 | beats | 5 | | 1 |

4 The games 'Three dicey numbers' and 'Abacus hundreds' (Resource sheets 6 and 7) can be adult-independent activities once they have been taught.

Plenary session

- *"Tom, come and make 204 on this abacus. What number would be 100 more than that?"*
- *"What was the largest number you made when you played 'Abacus hundreds'? What was the smallest number you made?"*
- *"Tell me something you learned today."*

Counting on and back

Assessment focus

- Can the children count on and back in 10's and 100's starting from any two- or three-digit number?

Resources

- Resource sheets 5, 9, 10 and 11
- large 100 squares and number lines
- several dice

With the whole class

- Give counting practice of some kind every day. For example, count in 10's/100's starting at 3 and going up to just above 100/1000. Ask *"What will be the first number we get to after 100/1000?"* You or a child should point to the numbers as you count along a large class number line or 100 square.

With the lower-achievers

With adult support

Choose from:

1 It is crucial that children struggling with counting receive as much adult support as possible. Even two minutes a day can help enormously. Create a different two-digit number with arrow cards each day and use that number to count in 1's, 10's, 100's and aim for 1000's by the end of the year.

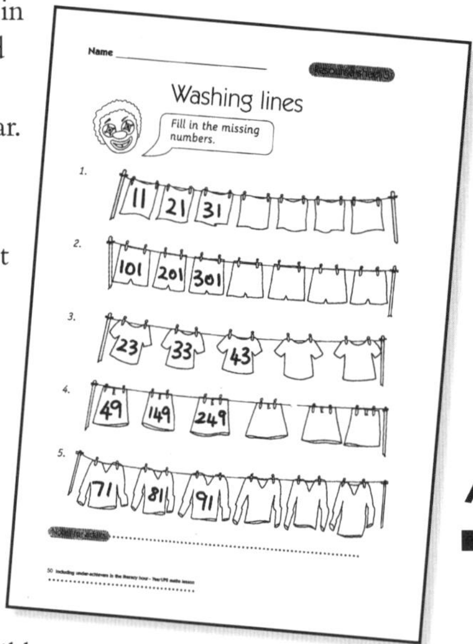

2 Write some starting numbers on Resource sheet 5, such as 3 in line 1 and 15 in line 2, and a rule to follow, such as counting up in 10's. The children fill in the missing numbers.

3 Make sure the children can relate counting in 10's or 100's to partitioning numbers. For example, with a starting number of 39 and a rule such as 'adding 100', the numbers are 139, 239, 339 and so on.

2	3	9	+ 100 is	3	3	9

4 Check that the children can name each number appropriately and continue to give practice with reading and writing numbers in words.

Teacher-independent activities

Choose from:

1 The children can play race games, such as throwing a dice and moving that many spaces along number lines and tracks. Resource sheets 10 and 11 can be used for this. You can cut up Resource sheet 11 to make a long number line. The children should record the numbers they land on.

2 Challenge them to record counting in a pattern, such as 10's or 100's, using Resource sheet 5.

3 Resource sheets 9 and 10 can be used to make a number track up to 1000 for display on the wall.

Plenary session

- *"If I'm counting in 100's and I just said 782, what is the next number?"*
  - *"What did you enjoy doing today?"*
  - *"Counting is very important, so ask someone at home to help you with it if you know you need more practice."*

Estimating and rounding

Assessment focus

- Can the children give sensible estimates of numbers of objects to 100, check count by grouping in 10's, and round two-digit numbers to the nearest 10 (then to the nearest 100)?

Resources

- Resource sheet 12
- lots of cubes/buttons, plastic jugs, arrow cards
- felt-tipped pens or mini stamps (Crayola, available in many supermarkets)
- A4 paper for making estimating sheets
- containers of sweets, biscuits and so on

With the whole class

- Before this session let children who need more support with counting make a series of estimating sheets using mini stamps or felt-tipped pens, printing between nine and

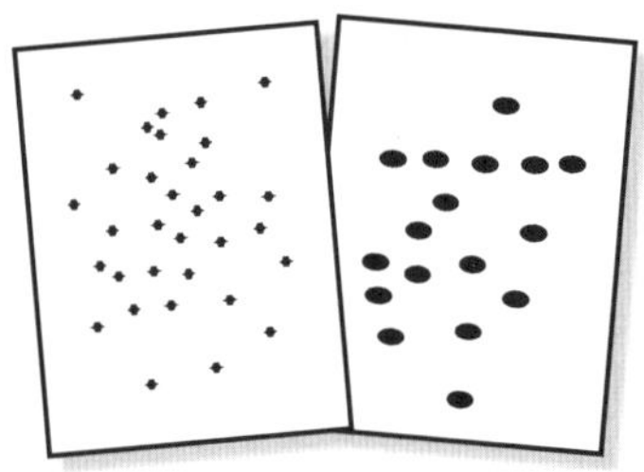

100 shapes on a page and writing the actual number of shapes on the back.
It doesn't matter too much if the count is slightly wrong, as the sheets are for estimating.

- Hold up the estimating sheets and ask for guesses. Write down the guesses to check later and mark them on the number line. Use the language of rounding as you do this activity – *"Jane guessed about 90. Look how close to 100 that is. We always round numbers that end in 5 up to the next 10."*
- Invite the children to take handfuls of cubes/buttons, make a guess then put the objects into jugs to count later.
- Make some of the numbers guessed with arrow cards. When you repeat the activity round these guesses to the nearest 10, again using arrow cards.

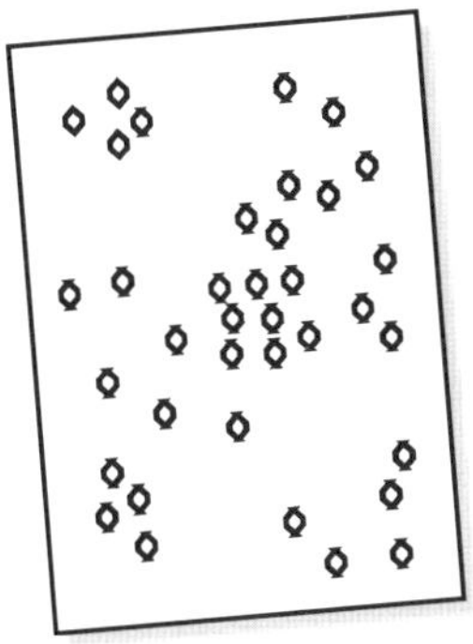

| 3 | 5 | rounds to | 4 | 0 |

"Salik guessed there are 35 and that is rounded up to 40."

With the lower-achievers

With adult support

Choose from:

1 Give repeated help with estimating items to about 10, 50 and 100, then counting by grouping in 10's. Observe – who is confident counting by grouping? Whose estimates are still way out? Vary the objects used, for example marbles in a jug, biscuits, raisins or dried pasta shapes in a packet.

2 Use number lines frequently to ensure the children can estimate the position of their number on the line and can round that number to the nearest 10 (then 100). It is crucial that estimating and rounding are secure as the children will need these skills for calculating.

Teacher-independent activities

Choose from:

1 Let the children estimate cubes in jugs, then group them in 10's to count totals. (Make sure they count in 2's and not in 1's to make groups of 10.) They should then make those numbers with arrow cards.

2 Use Resource sheet 12 to round to the nearest 10. You need to fill in numbers on the lower half of the sheet. Photocopy that lower half as many times as necessary if further practice is needed.

3 Sweets in a clear jar is always a popular and motivating activity to practise estimations. Let children record their guesses then count.

Plenary session

- *"Who was good at guessing today?"*
- *"Round 43 to the nearest 10. Do we go up or down when the unit digit is 3?"*
- *"How many biscuits were there? Is that enough for one each or do we need to break them in half?"*
- Play 'Grab'. Working together, pairs of children use all their four hands to take just one grab of cubes and then estimate how many they have picked up. Vary the game with items of different sizes.

Number sequences and reasoning

Overall learning objectives

- Recognise and extend number sequences, odd and even numbers, multiples of 2, 3, 4, 5 and 10.
- Reason about numbers.

Key words

predict	pattern
next	odd
rule	even

What's next?

Assessment focus

- Can the children recognise and extend number sequences, stating the rule?

Resources

- large number cards and small cards for the children (Resource sheets 13–19)
- Resource sheets 1, 5 and 20
- Blu-tack
- number lines and 100 squares

With the whole class

- Lay out large number cards in an easy sequence, such as 0, 2, 4, 6, 8 … and ask the children how they know the next one in the sequence. Answers should include *"It goes up in 2's."* or *"You add on 2 each time."*
- Go over the multiples of all the tables learned so far, sometimes going from a large number to smaller numbers.
- Counting in 10's and 100's as suggested in Chapter 1 needs to be repeated here.
- Make sure the children can add 10 to any number.

With the lower-achievers

With adult support

Choose from:

1 Show the children how to use the calculator constant to generate multiples. With most calculators this is 2 + = = = = to get multiples of 2. (With some it is 2 + + = = = =.)2 Play 'Will we hit 20?' Start a number pattern with cards, for example 0, 3, 6, 9, and ask *"Will we hit (use) 20?"* With 3's you won't. Say *"Start a pattern where we will hit 20."* Look for those children who can work this out quickly. Provide 100 squares or Resource sheet 1 for anyone who is struggling with this game. Ask questions such as *"How did you know that was the next number?"* and *"Tell us the rule for this pattern."*

3 Give the children practice in using 10's and 100's, counting forward and backward. Look out for children who need much more help with counting.

Teacher-independent activities

Choose from:

1 The children can work in pairs with small number cards making sequences of numbers. Blu-tack them to boards. Give them plenty of blank cards to continue the sequences. Keep number lines and 100 squares available.

2 Use Resource sheets 5 and 20 for recording sequences. You can fill in parts of the sheets and ask the child to fill in the blanks. Let them generate their own patterns as well as this often gives valuable insight into their understanding.

3 Some children might need to use the 100 square on Resource sheet 1 to go over the multiples learned so far.

Plenary session

- *"My number sequence has a 2 in it and a 6. What could my rule be?"*
- *"If all your numbers in your pattern ended in zero, what could your rule be?"*

Odd and even

Assessment focus

- Can the children recognise odd and even numbers?

Resources

- a large 100 square
- number cards and several copies of made cards: 'odd', 'even', '=', 'true' and 'false'
- Resource sheet 21
- calculators
- Blu-tack

With the whole class

- Chant the odd and even numbers using the 100 square and talk about how to recognise them by endings. Blu-tack two sets of cards on the board, one even and one odd set. Leave this on display and repeat the activity over a few days.
- Explore odd and even numbers with pairs of cubes. Odd numbers always have a cube sticking out. Even numbers are always in pairs.

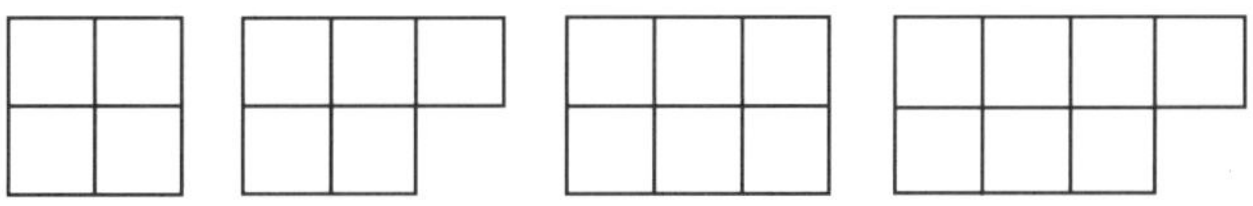

Make a series of cubes that children can continue on their own.

- Explore what happens when you add an odd number to an odd number. Display on the board 'odd + odd = even' and ask the children to find several ways of testing it out with a variety of numbers. Use calculators for larger numbers. Ask *"Are you really sure it will always be true even with larger numbers?"*

With the lower-achievers

With adult support

Choose from:

1 Give more help with reasoning about numbers, adding odd and even numbers, continuing by adding three then four numbers. Ideally you will end up with the children remembering what is true and what is false. Ask *"How could you test that out?", "Are you sure that will always be true even with big numbers?" and "How do you know you are right?"*

2 Make sure the children can identify odd and even numbers up to and over 100 .

Teacher-independent activities

Choose from:

1 Give the children calculators and ask them to explore adding odd and even numbers and testing out what is true and what is false.

2 Ask them to continue the sequence of cubes in pairs as shown in the whole-class starter and record their work on Resource sheet 21.

Plenary session

- *"Is this a true statement? All even numbers end in 2."*
- *"Even numbers can always be shared out equally between 3. True or false?"*

Multiples

Assessment focus

- Can the children recognise multiples of 2, 5 and 10 (then 3 and 4)?

Resources

- number cards of multiples of tables you want to use and sets of cards to 100
- Resource sheets 1 and 22
- a calculator or table square for checking

With the whole class

- Give repeated help with working out and memorising the number sequences of the times tables. Make links between them. For example, if you double the answers in the 2 times table you get the answers in the 4 times table.
- Chant around the circle counting in 1's. When you reach a multiple of the table you are practising everyone shouts the number. (Verbal number games can sometimes be stressful for

lower-achievers, so if everyone shouts out together, the fear of being wrong is reduced.)

- Introduce some of the games below.

With the lower-achievers

With adult support

Choose from:

1 Teach the children to play the game 'Sort it out'. Each child needs a copy of Resource sheet 22. Shuffle sets of cards up to 100 and place them face down on the table. Tell the children they are going to take a card and decide whether or not it is a multiple of the number you have chosen, say 10 (or whichever times table you want to work on) and whether or not it is odd or even. They take turns to take a card and place it on the sheet in the correct space. At the end, count the cards in each space. The player who has placed the most cards in the correct place is the winner.

Ask *"What do you notice?"* (There are no odd numbers that are multiples of 10.) *"Are multiples of 5 odd or even?"* (Both.)

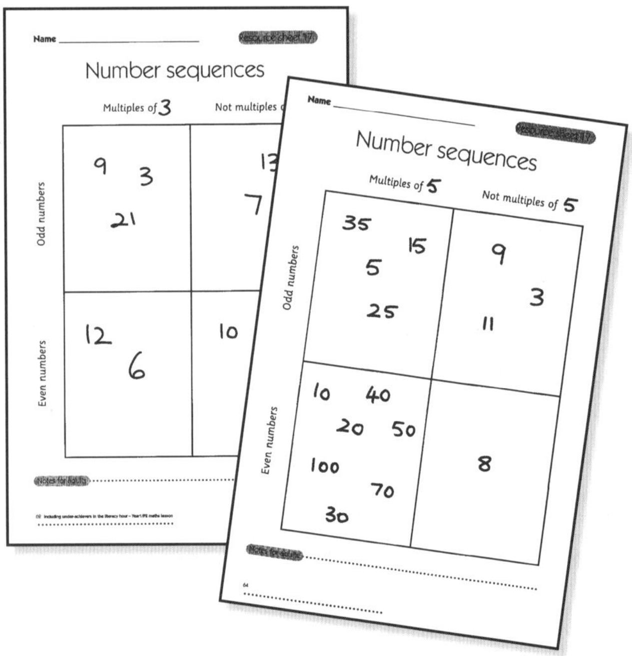

2 Play the 'Tables game'. Number cards are shuffled and placed face down on the table. The children take turns to turn over two numbers and if they are in the same times table, say 35 and 50, the child says *"These are both in the 5 times table"* and keeps that pair of cards. If the cards are not a pair they are replaced in exactly the same place. (Have a table square or calculator for checking.)

3 Play the 'Doubles game'. This game is played in the same way as the 'Tables game' but the children use cards of multiples of 2 and 4 and pair them up. So if a child finds 6 (3 x 2) she must find 12 and say *"I have 3 times 2 and this is 3 times 4."* Repeat this with the 5 and 10 times tables.

Teacher-independent activities

Choose from:

1 Ask the children to sort some number cards into sets using Venn diagrams.

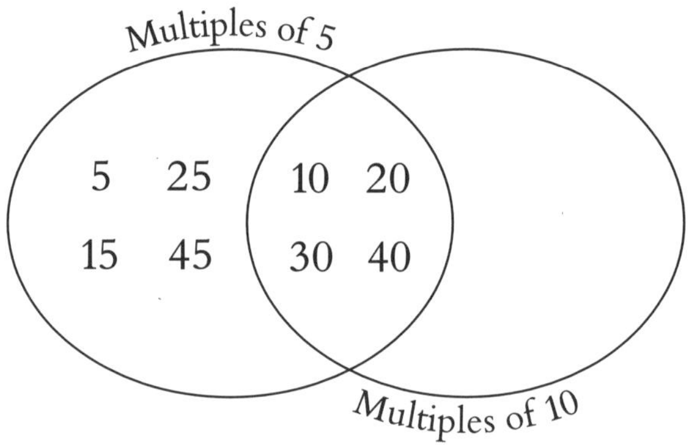

2 Using a 100 square, the children should colour yellow all the multiples of 2, put a blue circle around multiples of 5 and a green square around multiples of 10. Ask *"Which numbers are yellow, blue and green?"* (Multiples of 10.)

3 The three games described in the adult support section above can be teacher-independent activities once they have been taught.

Plenary session

- *"Tell me the different tables the number 6 is in."*
- *"45 is a multiple of 10. True or false?"*
- *"Is it true that if you double the answers in the 5 times table you get the answers in the 10 times table?"* (Yes.)
- *"If I know 5 times 10 is 50, tell me an easy way to work out 5 times 5."* (Halve 50.)

Addition and subtraction strategies

Overall learning objectives

- Add and subtract numbers mentally and on number lines, using a range of vocabulary.
- Use a range of calculating strategies for addition and subtraction, including adding and subtracting on a number line (then mentally), a 'near multiple of 10' to or from numbers up to 20 (then other two-digit numbers).

Key words

inverse	how many more
difference between	how many fewer
easy pairs	

Two-hand wizz

Assessment focus

- Can the children add two, three and four small numbers together mentally?

Resources

- Resource sheets 11, 23 and 24
- cubes and number cards

With the whole class

- Ask two children to come to the front. They hold their hands behind their backs and, at a given signal, bring out one/both of their hands holding up any number of fingers. Add the two/four numbers together. Check the answer by adding the numbers in a different order and also by subtracting the numbers in turn from the total. If your children are confident, they could be in teams and race to call out the correct total.
- Vary the game by letting the children hold up number cards, making sure you get numbers that total 10, such as 7 and 3, and so on.

With the lower-achievers

With adult support

Choose from:

1 Repeat the 'Two-hand wizz' game, first just with two children. Show them how to check their addition by adding in a different order.

2 Check that they can add mentally with small numbers and give plenty of practice to those who are still struggling with this (see number bonds in Chapter 4).

3 Using number cards, the children turn over two or three at a time and add them as quickly as they can in their heads. The first child to call out the correct answer wins the cards.

Teacher-independent activities

Choose from:

1 Fill in the number line on Resource sheet 11 or 23 to suit your children. For example, you could record the addition of two, then three, then four numbers.

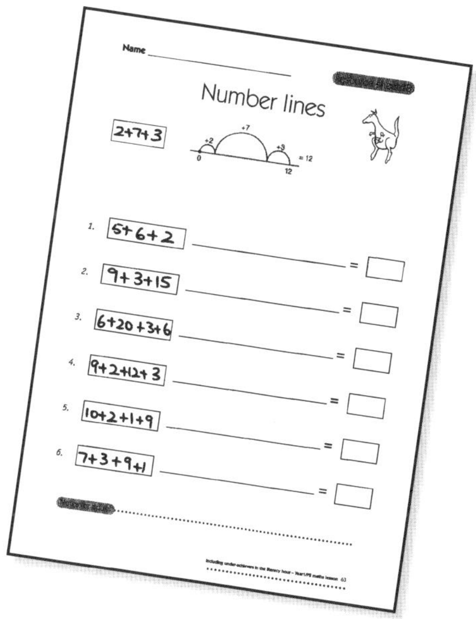

2 Each child needs a cube 'train' of 10 or 20 or to suit the child's ability and a copy of Resource sheet 24. Write the number of cubes worked with in the box at the top of the page. Explain to the children what to do. *"Split the cube train into two or three bits, for example 20 can split into 18 + 1 + 1 or 17 + 2 + 1."* Ask them to draw and

write their splits on the sheet. Repeat this on another day with a different picture from Resource sheets 69, 70 or 71 stuck at the top and a different number of cubes.

Plenary session

Play 'Two-hand wizz' again. Ask the children

- *"What must I add to 7 to get 20?"*
- *"What number when added to 10 makes 16?"*
- *"Add these numbers in your head, 3 and 4 and 5 and 6. How did you do it?"*

3 numbers game

Assessment focus

- Can the children use a range of language for addition and subtraction?

Resources

- cubes, board and pens
- Resource sheets 13, 14, 25 and 26
- paper for other words not on the sheet
- glue stick, Blu-tack, number lines

With the whole class

- Each child needs 'trains' of 10/15/20 cubes. Demonstrate splitting the 10 cubes into 3 and 7. Show them how 3 plus 7 is 10 and write this number sentence on the board, or Blu-tack the number and word cards on the board. Ask a child to tell you a different number sentence, still with 3, 7 and 10, but this time using different words, for example *"The difference between 7 and 10 is 3."* Read each number sentence carefully, involving the lower-achievers in this as it is ideal if they can read the words.

With the lower-achievers

With adult support

Choose from:

1 Using Resource sheets 13, 14 and 25, make number sentences where the children have to read the words and put the cards in an appropriate order to create a sentence. Make sure they can write sentences such as '7 is 3 less than 10' and '10 is 3 more than 7'. Do this with lots of different groups of three numbers.

2 Give the children plenty of time to use the words 'difference between' using cube 'trains' and number lines.

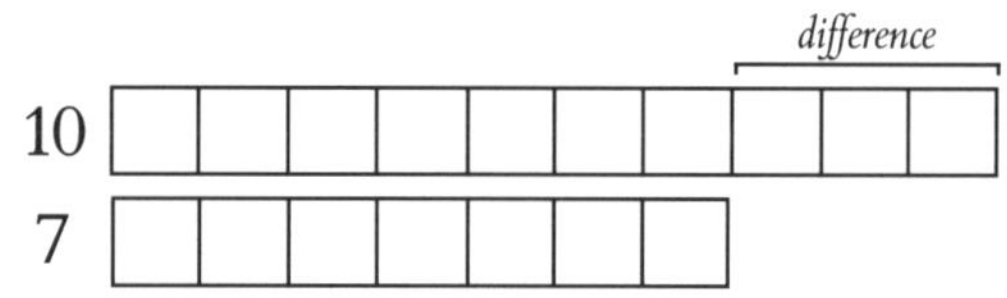

"The difference between 10 and 7 is 3."

Teacher-independent activities

Choose from:

1 Give the children copies of Resource sheet 25 and explain that they are to work in pairs to make as many number sentences as they can with three related numbers, such as 6, 9 and 15. Keep the sentences for display. Encourage them to think of their own words and write them down. Repeat the game with different numbers.

2 The children could record some sentences on Resource sheet 26. Say *"In each of the '3 numbers' games you always work with the same three numbers. This time you are working with numbers 6, 9 and 15."* It can be useful to ask them to make up their own three numbers to see if they have worked out that the numbers must be related.

Plenary session

- *"What could be my third number if I started with 8 and 3?"* (It could be 11, but some children might say that it could be 5.)
- *"The number in my head is 20 more than 17. What is my number?"*
- *"Tell me a word that could go in a subtraction sentence."*
- *"Which of these words are you unsure of?"*

Adding 9 by adding 10

Assessment focus

- Can the children add (and subtract) a near multiple of 10 on a number line?

Resources

- a large 100 square and number lines
- dice, two large pots, boxes, lots of cubes and counters
- spinners
- Resource sheets 1, 11, 23, 27, 28 and 29
- Blu-tack

With the whole class

- Make sure that all the children can add and subtract 10 to or from any number. Observe who needs more help with this. It is a priority to get addition and subtraction of 10 firmly established.
- Ask some children to the front to demonstrate how to add 10 on a number line, then hop back 1.

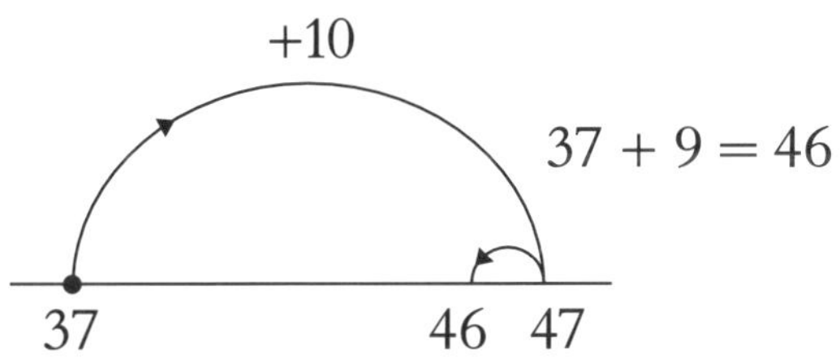

37 + 10 is 47. 37 + 9 is 46.

- Play a '4-in-a-row' game in teams (Resource sheet 27). Draw a copy of the game on the board, as shown top right. The children play in two teams with Blu-tack counters. The aim is to make four numbers in a row. (See the example at the top of the next column.)

Rules

i The children choose a number from the smaller box, then spin the spinner to see if they need to add 9 or 10. Then that team covers their answer.

ii The first team to make four in a row in any direction is the winner. They will see that it is a good strategy to spin the spinner first before they choose their number from the box in order to cover a number to block the line of the other team.

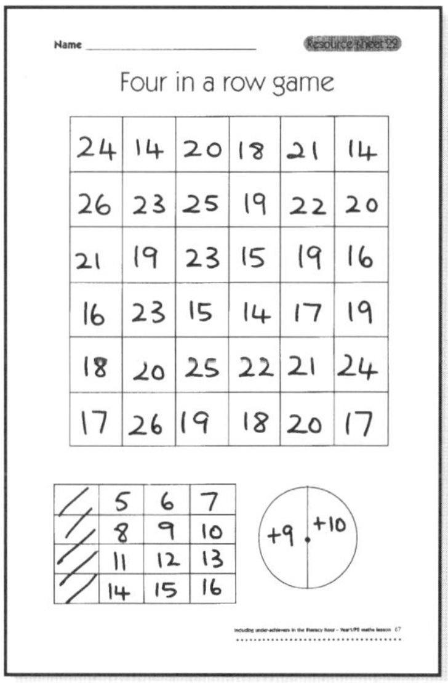

Name

Four in a row game

24	14	20	18	21	14
26	23	25	19	22	20
21	19	23	15	19	16
16	23	15	14	17	19
18	20	25	22	21	24
17	26	19	18	20	17

/	5	6	7
/	8	9	10
/	11	12	13
/	14	15	16

+9 +10

With the lower-achievers

With adult support

Choose from:

1 Play 'What's in the pot?' using two copies of Resource sheet 28 filled in something like the example here.

Rules

i You need 2–6 players in two teams, two pots, two dice, two counters, cubes, long write-on number lines up to at least 100 (Resource sheet 11) and plenty of blank number lines to add on as the numbers can go up to and beyond 200.

ii Each team puts their counter on the 'Start' on their own sheet. The first player for each team throws the dice and moves that many spaces. They follow the instruction on the space they land on. For example, if the space says 'add 4' they put that many cubes in their pot. They should draw their calculation on their number line each time. Then the next player throws the dice, moves that many spaces and follows the instruction. At the end, the final number arrived at on the number line should equal the number of cubes in the pot! Supervise the jumps the children make on the number line, checking that they add near multiples of 10 appropriately.

iii The winning team reaches the highest (or lowest) total – or is the team that ends up with the same number of cubes in their pot as they reach on their number line!

iv Adapt the game for different children. You could use just multiples of 10 to give the children further practice, for example + 50, + 10, + 30, –40 and so on along the track.

2 Using Resource sheet 29 and blank number lines, teach the children a simple race game. Fill it in something like the example below.

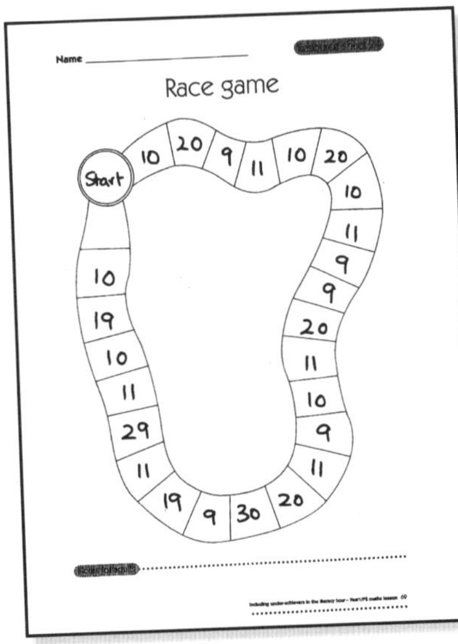

Rules

i Choose a target number, such as 20. The children take turns to throw a dice and move their counter around the track. The idea of the game is to get back to the start as close to the target number as possible.

ii Everyone starts with 20 on their number line. They throw a dice and land on that square on the track. They then have to decide whether they will add the number they have landed on to the 20 on their number line or subtract it.

iii The next time they land on a square they must make the same decision but they should be trying all the time to keep as close to 20 as possible. Observe who is adding mentally with 9 and 11. Who can add accurately on a number line?

Teacher-independent activities

Choose from:

1 These children can play '4-in-a-row' using Resource sheet 27 with numbers filled in to suit their abilities. Eventually let them make up their own game, gradually making the spinner more complex. For example, divide the spinner into four and have + 19, – 19, + 11, – 11.

2 Prepare number lines on Resource sheet 23 with calculations such as 34 + 9 or to suit your children. Remind them they have to draw the calculation on the number line.

3 For those still needing more help with adding 10, (or 20 and 30 and so on) this is best done with a 100 square (Resource sheet 1).

Plenary session

- *"Make a number line in your head and add 38 and 9."*
- *"Which other numbers would be easy to add like 9?"* (19, 29, 39 and so on.)
- *"What would you like to practise again?"*

Larger number first

Assessment focus

- Can the children put the larger number first when adding numbers?

Resources

- number cards, coloured pencils, a 1–6 dice
- Resource sheets 9 and 23

With the whole class

- Demonstrate how it is very slow to add in 1's when the number is large, such as 3 + 51. Say *"If you add in 1's that would be very slow. How could you do 3 + 51 much more quickly?"* Establish that addition can be done in any order and therefore putting the larger number first makes it quicker. Show the children pairs of numbers on number cards, for example 5 and 17, and ask a lower-achiever to the front to swap them round physically.
- Show the children how to record activity 3 on Resource sheet 9. (This is described opposite in the Teacher-independent activities section.)

With the lower-achievers

With adult support

Choose from:

1 Play 'In my head' with number cards. Put several larger number cards on the table face up and several smaller cards face down. The children take turns to select a larger number and turn over a smaller number to add on. They say something like *"If I put 43 in my head then I can add on 3 on my fingers."* Observe – is anyone wanting to start from 1 each time? If so, you will need to keep practising *"Put the first number in your head and count on."*

2 Do some addition with coins starting with 20p or 50p and adding on a few pence.

Teacher-independent activities

Choose from:

1 Give out the number cards 1–5 and 20–25 to pairs of children according to their ability. They have to take two numbers, put them into pairs with the larger number first then work together to do each calculation. They can record these on Resource sheet 23, using the number lines.

2 Fill in Resource sheet 23 with some calculations where the smaller number is first, for example 2 + 45. The children have to reverse the order and write the calculation on the number lines.

3 Using Resource sheet 9, give each pair of children about six starting numbers, such as 34, 56, 89 and six different coloured pencils. They have to throw a dice to find the number to add to those starting numbers, colouring in the start number in one colour and using the same colour to show the spaces they jumped on to get to the total, such as 34 + 5. They should then change to another start number and another colour.

Plenary session

- *"Why do we put the larger number first when we add?"* (It's quicker, but note that there are times when we wouldn't add in that way, for example when adding 25 and 75. This is often a known fact as is 30 add 31. We wouldn't bother to reverse near doubles.)

Two hop hundreds

Assessment focus

- Can the children find pairs of numbers to 100?

Resources

- Resource sheets 23, 30 and 71

With the whole class

- Ask the children to help you make pairs of numbers to 20. *"What goes with 7 to make 20?"* Gradually build up through 30, 40, 50 then 100, leaving pairs of numbers on display and going over them with the lower-achievers for two minutes every day for several weeks.

- Extend this work to pairs of multiples of 50 that make 1000, for example 950 and 50.

With the lower-achievers

With adult support

Choose from:

1 Play this memory game to help the children to memorise pairs of numbers to 20/50/100 and so on. Select some number cards that make pairs to 100 and place them face down. The children turn over two cards at a time. If they add up to 100 the player keeps that pair. If not, they are put back in exactly the same place.

2 Use Resource sheet 23 to make sure the children can draw hops on a number line and tell you what they do for a range of different calculations. Ask *"Can you make a number line in your head and add 19 and 7?"*

Teacher-independent activities

Choose from:

1 The children can play a game called 'Two hop twenties'. Working together in pairs using Resource sheet 23, they make number pairs, for example to 20, taking 2 hops on the number line to get to 20.

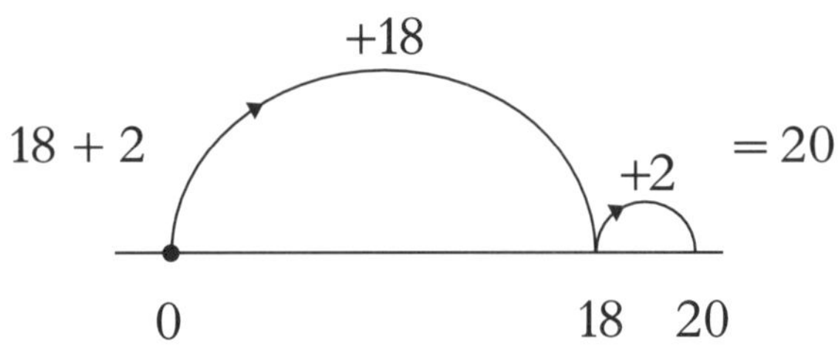

2 Using 0–100 number cards, the children could work in small groups to make pairs to 100 and Blu-tack them onto a board.

3 They can do the same activity but this time using coins to make piles of coins that together make a pound, for example 25p and 75p or 59p and 41p.

4 Write numbers into the dinosaurs on Resource sheet 30 so that the children can find all the pairs of numbers that add up to 20 (or another number of your choice).

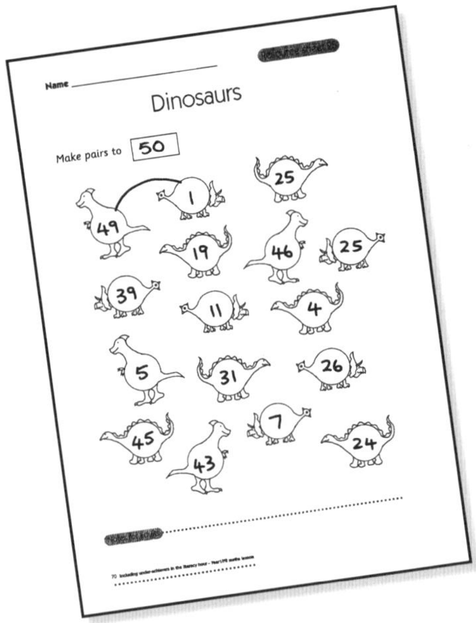

Plenary session

- *"Tell me a pair of numbers that make 100"*
- *"What did you enjoy doing most in maths today?"*
- *"What are some important things about today's lesson that you want to remember?"*

Easy adding

Assessment focus

- Can the children find easy pairs of numbers when adding?

Resources

- Resource sheets 31 and 71
- number cards and money cards with amounts such as 15p and 25p
- coins
- a number line

With the whole class

- Add up amounts of money making easy pairs to 10p, 20p and 30p, for example 15p + 5p = 20p. Display the lists of money on Resource sheet 31 and ask some children to come to the front to find easy pairs.

With the lower-achievers

With adult support

Choose from:

1 Create other bills to add up with the children introducing column addition in a real context. Observe who is making sense of 'easy pairs'. Who needs help with number bonds to 10 and 20? (See Chapter 4.)

$$\begin{array}{r} 17+ \\ 3 \\ \hline 20 \\ \hline \end{array}$$

2 Using real or plastic coins, select an amount of money, for example 17p, and ask the children to make it up to 20p, counting on in pennies if necessary. Record this on a number line.

3 Adapt the last activity to giving change, counting up on the number line. For example, *"The juice costs 24 pence. How much change from 50 pence?"*

Teacher-independent activities

Choose from:

1 Ask the children to use number cards or money cards to make pairs that equal a multiple of 10.

2 Resource sheet 31 could be done in pairs using colours to show which numbers have been linked together to make 'easy pairs'.

3 Create more money sheets like Resource sheet 31 using Resource sheet 71. For example, cut out four to six items and price them to suit your children, such as juice priced at 17p and a lolly priced at 3p.

Plenary session

- *"Why do we find easy pairs of numbers to add?"* (Because it makes adding quicker.)
- *"I'm going to say a number and I want you to call out the number that makes my number up to 100."*

Chapter 4

Linking addition and subtraction

Overall learning objectives

- Understand that subtraction is the inverse of addition and state the subtraction corresponding to a given addition and vice versa.
- Know the number bonds of all numbers to 10 and 20.

Key words

inverse	total
two-digit numbers	minus
sum	count on/back

How many are hiding?

Assessment focus

- Can the children work out a missing number in addition and subtraction sentences?

Resources

- cubes or play people
- a box or cloth
- board and pens
- Resource sheets 24, 32, 69, 70 and 71

With the whole class

- Do this activity several times with different numbers of cubes/play people. If you can't work on the floor with the children, use a cube 'train' of ten cubes and hide the cubes behind your back. With the ten items, set up a story situation. For example, *"There are 10 people on the bus…"* (put 10 play people out of sight in a box) or *"There are 10 clowns inside the big top…"* (put 10 cubes under the cloth) and so on. Say *"3 of them get off the bus."* (Show the 3 items.) *"How many are still on the bus?"* Work this out with fingers if necessary. Repeat this working through all the number bonds of 10. Ask some children up to the board to write up a number sentence, for example 10 – 7 = 3, 7 + 3 = 10, 10 – 3 = 7 and 10 – 7 = 3. To give the children experience of empty box problems, include statements such as $10 - \square = 3$.

With the lower-achievers

With adult support

Choose from:

1 Repeat the activity done in the whole-class starter session, letting the children record the number sentences including empty boxes. Read the sentences with individuals. Observe – who can read an empty box problem appropriately?

2 Use a range of different equations. It can be helpful to use a pair of balance scales.

$15 + 5 = \square + 6$ $\qquad \bigcirc + \triangle = 19 + 1$

$\square = 17 + 3$ $\qquad \square - \square = 20$

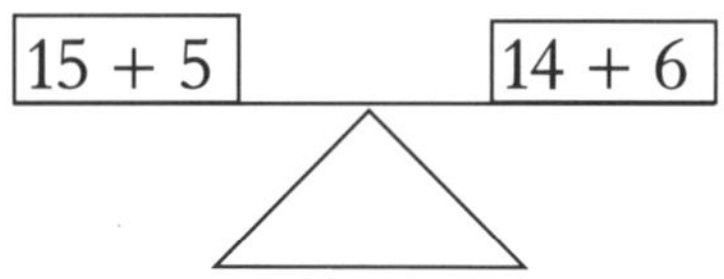

Give the children plenty of experience in reading a variety of equations.

Teacher-independent activities

Choose from:

1 The children can play the game they did in the whole-class starter activity, working in pairs and taking turns to hide the cubes. Ask them to record their work in pictures, showing the number of people on the bus and the number off the bus (or clowns in the big top and the number out of the big top). Encourage them to be systematic. For example, with number 9, start with 9 in and 0 out, 8 in and 1 out and so on. When the children are confident with being systematic they can use Resource sheet 32 to record, using the space on the right to continue the sequence of numbers. Reuse the sheet with a different number and one of the pictures from Resource sheets 69–71.

3 Resource sheet 24 can also be used to record number bonds.

Plenary session

- *"Someone come and write on the board an equation with an empty box in it."*
- *"What do empty boxes stand for?"*
- *"Who can read this equation?"*
- *"What does the = sign mean?"* (balances, equals, makes)

Number bonds

Assessment focus

- Can the children remember number bonds to 10 and up to 20?

Resources

- Resource sheets 13, 14, 24, 30 and 32
- different coloured pens and some card

With the whole class

- Play 'Ping pong'. Start with a target of 10. You say a number, for example 7, and the children must reply with 3, making your number up to 10. Every so often, instead of saying a number say *"Ping"* and the children have to reply *"Pong"*. This makes the game fun and keeps them on their toes! Repeat this every day for several weeks until all the children can do it to 10 and some can do numbers to 20.

With the lower-achievers

With adult support

Choose from:

1 Make a chart of the children's names and the number bonds that they know by heart. Work with them to fill in this chart so that they know they are learning all the number bonds from 5 up to 20. Pair the children so that a child working on number bonds to 7 is working with the child working on 17.

2 Play 'Ping pong' again and use the paired number cards as above. Ask *"If you are working with number 15, what is the number pair that goes with 14?"* (1)

3 Using Resource sheet 24 or 32, give the children a number and ask them to write all the number bonds for that number systematically on the page. Some of them might need to use cubes.

Teacher-independent activities

Choose from:

1 Copy Resource sheets 13 and 14 several times onto coloured card, for example numbers from 1 to 10 on yellow card, from 1 to 12 on red card and from 1 to 15 on blue card. Using the different coloured number cards, cut them up and ask the children to take one colour and work in pairs to write the 'Ping pong' pair on the back of the card in pencil. So, for 10 the card with 5 on it has 5 written on the back and 7 has 3 on the back. (Check their work before the numbers are written in pen. The numbers on the back need to be written in a different colour from those on the front.)

2 Use the sets of number cards to play a game in pairs, with one colour of cards to each pair. Lay the cards so that the black numbers are facing up. In turns, the children challenge their partners to find a particular number, for example *"Find 7."* (This is on the back of 3.)

3 Let the children use Resource sheet 30 to record pairs. They could write in their own numbers, making sure they make pairs to 20 then draw lines between the pairs.

Plenary session

- Invite one child to come to the front and hold up a large number card. This is the target number. The child gives the 'Ping pong' number to the class. The class must call out (or whisper) the number that makes this up to the target number. If a child is holding up 8, but forgets that he must only call out numbers up to 8, ask what you could do.
- You can adapt the game for subtractions as well. If 8 is on the card and 10 is called out, the reply is *"Take away 2."*

If I know this, what else do I know?

Assessment focus

- Can the children use corresponding addition and subtraction facts?

Resources

- number cards and '+', '–' and '=' cards
- small pieces of paper and Blu-tack
- number lines and small boards
- Resource sheets 33, 69–71

With the whole class

- Ask the children *"If I know that 9 add 3 is 12, what else do I know?"* Write up their suggestions, for example 3 + 9 is 12.
- Make the point that addition can be done in any order but subtraction can't. 12 – 9 is 3 and 12 – 3 is 9. Provide several examples, asking the children to the front to write the number sentences on the board and also to draw the sentences as hops on a number line. (Link this to Chapter 3 and the variety of language used in the '3 numbers' game.) Sometimes start with a subtraction, for example *"If I know 15 subtract 5 is 10, what else do I know?"*

With the lower-achievers

With adult support

Choose from:

1 Let the children use Resource sheet 33 for recording their own numbers. Reuse it as many times as you like using a different picture from Resource sheets 69–71.

2 Work orally with the children, using a range of numbers and letting them be the person who asks *"If I know… what else do I know?"* Make the point that if they learn their number bonds really well, that helps to make lots more 'sums'.

3 Make a display of familiar number sentences and other 'sums' that can be made from them and go over these several times and at mental maths time.

Teacher-independent activities

Choose from:

1 Give the children small number cards along with '+', '–' and '=' cards. Ask them to create number sentences and stick them down with Blu-tack in families.

2 If they are familiar with recording on number lines, they can make drawings of each statement on number lines.

Plenary session

- *"Tell me a number bond you know and I will tell you what I can work out from what you say."* (For example, the child might say *"20 – 17 = 3"* and you will say *"So 20 – 3 = 17"*.)
- *"There is something you can do with addition that you can't do with subtraction. What is it?"* (Reverse the calculation.)
- *"Are you learning lots of things by heart in maths? What could help you with that at home?"*
- *"Jason's target this term was to learn all the number bonds to 20 and now he knows them all!"*

Stepping back and forward

Assessment focus

- Can the children understand that if you count back, then forward the same number, you end up where you started?

Resources

- a large floor number line to about 20
- board and pens
- Resource sheet 34

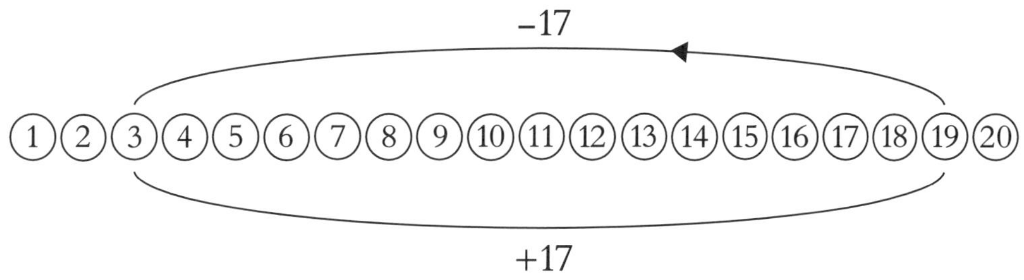

With the whole class

- Stand around the large floor number line or track, or make a simple track by writing numbers on bits of scrap paper. Use children to stand on the track. For example, *"Jessie, stand on 20. How many steps to get to 3?"* Count with the child as she steps and record the number on the board. *"Now start on 3 and predict how many steps you need to get back to 20."* Then show them how to draw the steps on a wall number line or track, as shown above.
- Stress that whether you count on or back, it is the same number each time and link this to the '3 numbers' game in Chapter 3.
- Introduce Resource sheet 34. Show the children how adding a number on, then taking away or counting back that same number means you get back to where you started.

With the lower-achievers

With adult support

Choose from:

1 *"Tell me what happens to the number 17 when you add 3 then take away the 3 again."* Explore the children's language of this activity. Observe – who has clearly understood the learning objective?

2 Work with the children around a number line or track and mentally add and subtract small numbers that relate to the number bonds they have been learning. Observe – who is mentally adding and subtracting securely?

3 Extend the activity to finding the differences between a pair of larger numbers by counting on or back. So, to find 45 – 18 you can do that by counting on from 18 on a number line. 18 + 2 is 20, add 20 is 40, add 5 is 45 so 45 – 18 is 27.

Teacher-independent activities

Choose from:

1 Let the group work around the floor number track, recording what they do in their own way.

2 Ask them to use Resource sheet 34 to show forward and back steps.

Plenary session

- *"If I start on 4 and step forward 10, where would I land? If I then step back only 9, where would I be?"*
- *"What happens to a number if you take away 36 then add the 36 back on again?"*
- *"What have you learned today about how addition and subtraction are linked?"*

Adding two-digit numbers

Assessment focus

- Can the children add two two-digit numbers?

Resources

- place value/arrow cards
- number lines
- Resource sheets 23 (a blank copy) and 35
- a large 100 square
- number cards 1–10 or a pack of playing cards

With the whole class

- Make sure that all the children can add multiples of 10, for example 20 + 30. Do this as hops along a number line as well as on a large 100 square. Remind the children to always check what they do.
- Then give plenty of practice with adding a multiple of 10 to any number on a 100 square, for example 14 + 20. Record these on a number line.

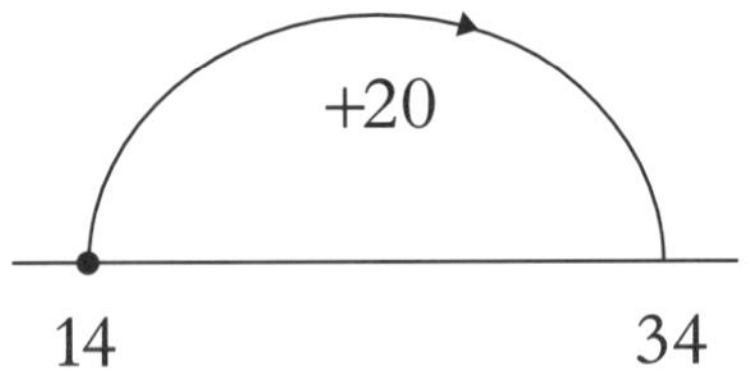

- Go over partitioning (see Chapter 1) into 10s and units.

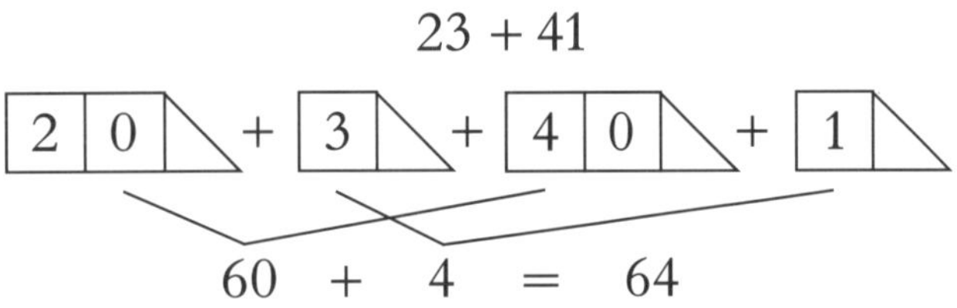

- Also record these additions on number lines. Show the children how to check their work.

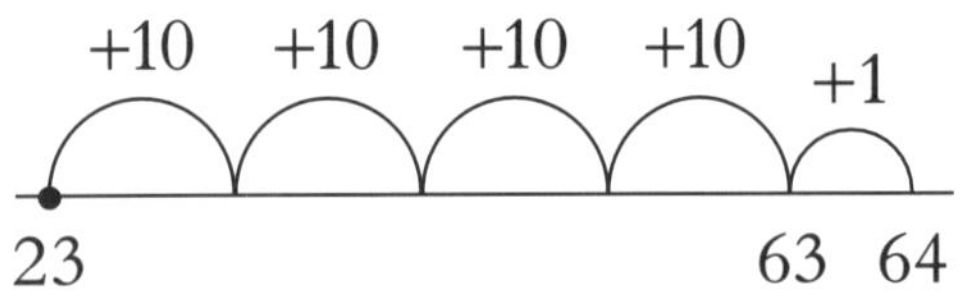

With the lower-achievers

With adult support

Choose from:

1 Give more practice with each aspect of the whole-class activities.

2 When the children are ready, show them how to record their additions on Resource sheet 23, linking this to their mental methods.

3 To give more experience in adding up to 100 show the children how to play the '100' game.

Rules

i Deal out a pack of playing cards so that each player has about six cards if you want the game to last a whole session. All picture cards can be 10 and ace can be 1.

ii The first player puts down a card, say 6. The second player puts down a card, say 5. That player adds 6 and 5 and everyone checks and writes down 11. (If you want, get the children to write the whole sum so that you can check when the game is played independently.)

iii Then the next player puts down a card, say 3, and adds that to 11. Everyone writes down 14. This goes on until a player makes 100 or above. That player is the winner.

Teacher-independent activities

Choose from:

1 Play the '100' game (see above).

2 Play the '24' game.

Rules

Put cards face up on the table. Players take turns to take one card trying to make a total of 24 eventually.

3 Play the '21 game'.

Rules

Deal out a pack of cards. In turn, players put one card face up in the middle trying to make tricks of 21. They play a card each time and if they can't make a trick, they start again. When they have made a trick of 21, they put it face down on the table and don't use those cards again. The winner has the most tricks of 21.

4 Use Resource sheet 23 when children are able to work more independently, for example to do calculations such as 35 + 21.

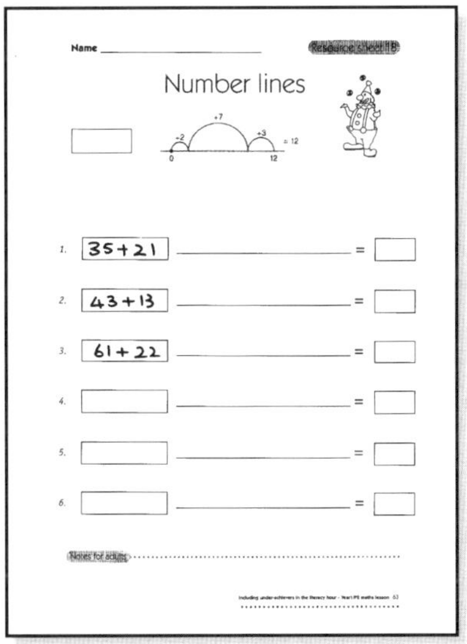

5 Use Resource sheet 35 to assess whether the children can do a calculation in two different ways and can understand how to check their work.

Plenary session

- Ask the children to work in pairs to do a calculation in two different ways, recording what they do on paper. Some of them can then come to the front to explain what they did and show their methods on the board. Ask *"How did you check what you did to be sure you're right?"*
- *"Who can add two two-digit numbers using a number line?"*
- *"What is your favourite way of doing that, Dean?"*
- *"So we can do additions and subtractions on a number line like this."*

Multiplication and division

Overall learning objectives

- Know by heart all the multiplication facts for the 2, 5 and 10 times tables (then the 3 and 4 times tables) and relate these to division facts.
- Use the language of multiplication and division and finding remainders.

Key words

lots of
groups of
arrays
row
column
inverse
times as big

Lots of

Assessment focus

- Can the children describe arrays using the language of multiplication?

Resources

- Resource sheets 27, 36, 37, 38, 69, 70 and 71
- large number lines
- dice, cubes, scissors and glue

With the whole class

- Start with saying some tables and counting in 2's, 5's and 10's. While they are sitting at their tables, give each child a pile of cubes, for example 12, and ask them to put them into equal groups, such as 3 lots of 4. They should then change the piles around so that there are 4 lots of 3. Make arrays. (See Resource sheet 36.) *"I've made 3 rows of 4."*
- Give the children plenty of memory practice, for example asking the same multiplication fact several times over a period of a few minutes. *"What is 3 x 2, 5 x 2, 1 x 2, 3 x 2, 0 x 2, 3 x 2?"* and so on.
- Demonstrate repeated addition on large number lines and link this to arrays. For example, *"We can make 5 hops of 2 and that is 2 + 2 + 2 + 2 + 2 or 5 rows of 2".*

With the lower-achievers

With adult support

Choose from:

1 The children can play the 'Lots of' game using two dice or spinners and plenty of cubes.

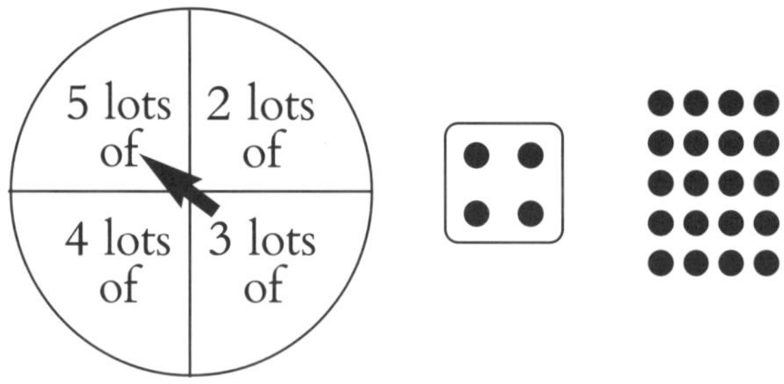

"I need to put out 5 lots of 4."

2 Extend the game by letting the children draw some of the arrays on squared paper and name them. Have copies of the tables on the wall near by. Go over the multiplication facts several times so that they have a chance to memorise them.

3 Ask the children to complete Resource sheet 37, using the number of cubes to suit each child, for example 12 cubes, such as on the sheet shown below. (Don't forget 1 row of 12.) The children make their arrays and either draw them or cut them out of 1cm squared paper and stick them on. Use the sheet again with a different number of cubes and a different picture from Resource sheets 69–71.

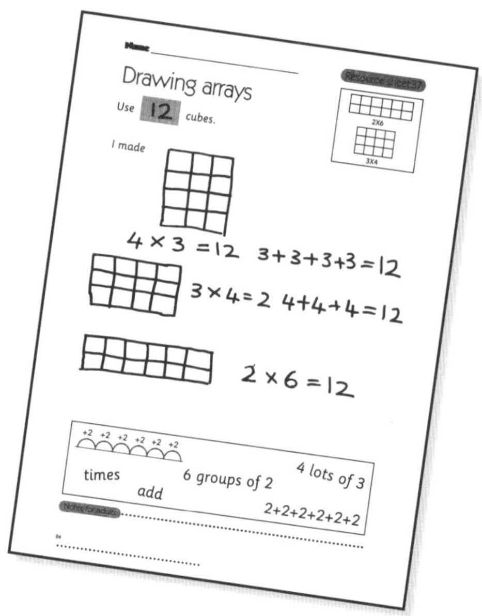

Good numbers are 18, 24, 36, 48. Some numbers give fewer arrays than others, for example 21 is only 7 x 3, 3 x 7 and 21 x 1.

4 The children could make their cubes into arrays and describe them to you. Observe – who is still putting out 3 lots of 4 as 3 and another 4? Who is using the language of 'lots of' appropriately? Ask the children to explain. *"Tell me about your array"* and *"Is Damien's array 4 lots of 5? How many altogether? Let's add them, 5 and 5 and 5 and 5. Tell me a quicker way to write that."* (There are pictures of arrays of 18 and 24 on Resource sheet 69 for you to make up other activity sheets.)

Teacher-independent activities

Choose from:

1 Ask the children to use Resource sheet 36 to make and describe arrays.

2 The children can continue the whole-class starter activity using cubes to make arrays and then copy them onto squared paper or onto Resource sheet 38. They can then cut these out and stick them onto paper for a display.

3 Ask the children to make some '4-in-a-row' games to practise. Use Resource sheet 27, white out the line on the spinner and add in a multiplication sign and any numbers you want to use.

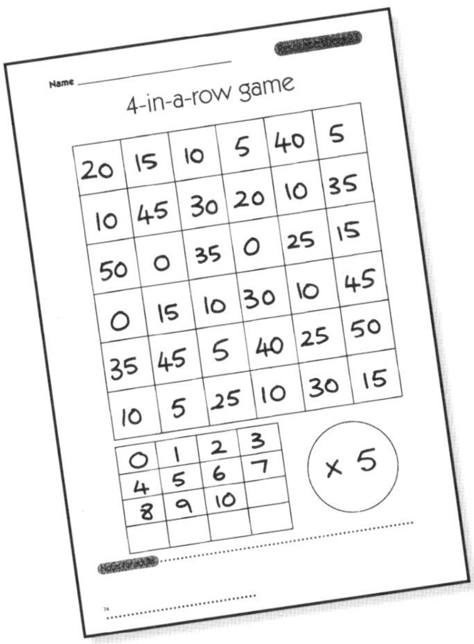

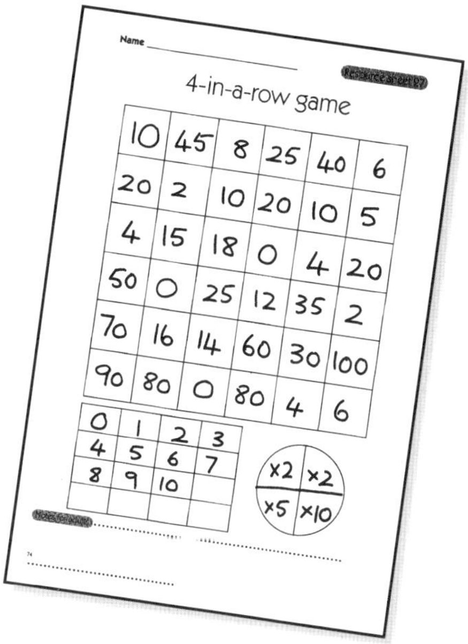

You can make the game have more of an element of luck by keeping the spinner space as a spinner and using more than one times table.

Plenary session

- Chant the tables and just the multiples, for example 2, 4, 6, 8 … Show some arrays on the board and ask the children to describe them and come to the front and write either multiple additions or multiplication calculations.
- Ask *"Who can say a table by heart with a partner?"*

Repeated subtraction on a number line

Assessment focus

- Can the children do repeated subtraction on a number line?

Resources

- large number lines
- Resource sheets 11, 39 and 40
- cubes

With the whole class

- Tell division stories that are about equal sharing. For example, *"There are 15 buns. How many each if we share them between 3 children?"* Also tell stories where the size of the group is decided and sometimes this means that some people get nothing! For example, *"The teacher has 14 apples. To how many children can she give 2 apples?"* Seven children can have 2 each, but if there were 10 children, 3 of them get nothing! It is these 'grouping' stories that are worked out by repeated subtraction on the number line.
- Demonstrate repeated subtraction on a number line, making it clear that this 'hopping back' is the opposite of multiple adding. That is why division 'undoes' multiplication. Introduce Resource sheet 39.

With the lower-achievers

With adult support

Choose from:

1 Give the children 'grouping' stories and ask them to represent these on number lines. For example, *"Tina has 20 sweets. To how many of her friends can she give 5?"* Clarify the language pattern of a grouping story compared with an equal sharing story. (There is almost always a 'to how many…?' or 'how many children can have…?' phrase.)

2 Use Resource sheet 40 to record grouping stories.

3 Some children might benefit from doing the hopping back by having long 'trains' of cubes and breaking off the size of the hop each time. *"I had 12 cubes and I broke off 3 so now there are 9 left."*

Teacher-independent activities

Choose from:

1 The children need to practise hopping back on number lines, as they have to start at the end and go backwards. They could practise this on Resource sheet 11. Provide calculations for them to do, such as *"Start on 20 and hop back in 5's."*

2 The children count the hops they make on Resource sheet 11. If it is appropriate, give them tables squares so that they can check their divisions with multiplication.

Plenary session

- *"Someone tell us a grouping story about sweets, where the children are going to get just 2 each, and some children might not get any."*
- *"Tell me something you learned today that is new knowledge for you."*

Dividing and remainders

Assessment focus

- Can the children divide into equal groups and also work out remainders?

Resources

- cubes
- large tables square or tables written out on the wall
- Resource sheets 41, 42, 43 and 44
- number cards

With the whole class

- Do some equal sharing, making sure all the children can share with a '1 for you and 1 for me' technique and explain what they have done using phrases such as 'equal groups', 'they get 3 each' and 'there is 1 left over'. Try to give the children practical experience of this by letting them work at their desks with cubes. Relate what you do to a large table square and/or copies of tables written out on the wall.
- Work on language such as *"I know 12 will share out into 3 groups because it is in the 3 times table."*
- Demonstrate how to use Resource sheet 41, where numbers are shared equally.

With the lower-achievers

With adult support

Choose from:

1 Teach the 'Share the biscuits' game with the children in two teams. You need one copy of Resource sheet 44 and number cards 5–15 with 7, 11 and 13 removed (or numbers to suit your children).

Rules

One team turns over a card, say 15, and they must decide which number of plates they will choose to share out the biscuits equally. So, with 15 they might choose 3 or 5 plates. They physically share out some cubes and if they have none left over, they win a point. If they end up with a remainder they can have one more chance to find a better number of plates.

2 Give the children challenging numbers for some division calculations. Observe – who is sharing with confidence and who makes equal groups without having to be reminded?

3 Teach them the 'Grab for gold' game on Resource sheet 43. They take turns to be the pirate who grabs cubes from a treasure chest just using one (or two) hands. They count their cubes, put the number in the first box, then share them out between two (for questions 1 to 4) and write the remainder. Each player wins the number of points of their remainders. Let other children check using a calculator as this gives further practice with the division sign and raises the question of decimal remainders.

Teacher-independent activities

Choose from:

1 Resource sheet 41 has very simple numbers that divide out equally with no remainders. Use Resource sheet 42 for more difficult numbers and more practice.

2 Working with number cards (all even numbers for those who need a very simple activity), the children can work in pairs to share out the number of cubes on the card (always into two groups for those who need that practice) and write and draw what they find out in their own way.

3 Extend activity 2 with a wider range of number cards and let the children choose to divide between 2 or 3 (or extend to 4 and 5 as well, using Resource sheet 44).

Plenary session

- *"My starting number is 15 and I share out equally but have a remainder of 1. How many people did I share between?"*
- Ask questions such as *"What is wrong with this? 7 divided by 2 is 2, remainder 3."*

Tables and division

Assessment focus

- Can the children derive quickly division facts that relate to multiplication tables?

Resources

- a large multiplication square or large copies of times tables displayed on the wall
- cubes
- flash cards of multiplications ('6 x 3' and so on)
- number cards of multiples of tables you are concentrating on
- Resource sheets 1, 38 and 44
- card and Blu-tack

With the whole class

- Count together in 2's, 3's, 4's, 5's and 10's as daily practice for several weeks.
- Go over a few multiplication facts and display them. The children can come up and Blu-tack the answers to the questions. For example, *"If I know that 7 times 2 is 14, what do I know about dividing 14 into 2 groups or 7 groups?"* Make a display of the calculations. You could set this out in the same way as on Resource sheet 33 – *'If I know…'* Make it clear that learning tables helps you to do divisions quickly.
- Play a game where you give out the multiplication flash cards and the answers. Set a timer to ring in one minute and see if the children can find their partner(s) and sit down. (For example, '6 x 3' finds '18'.)

With the lower-achievers

With adult support

Choose from:

1 Build up a multiplication chart for each child for the times tables 2, 3, 4, 5 and 10. You can do this on 1cm paper or use the 2cm squares on Resource sheet 38. The children should write in pencil on the chart at first and then check to see if they are right before they go over in pen. Mount their finished charts onto scrap card. Now make an L shape out of card for each child which they can use with their chart to find both multiplication and division answers, as in the example below.

X	1	2	3	4	5			8	9	10
2										
3										
4					20					

By placing the 'L' shape with the 20 in the right angle of the 'L' it can be seen that 20 ÷ 5 is 4 and 20 ÷ 4 is 5.

2 Show the children how to use the chart and the 'L' shape by using some number cards. For example, show them 30 and ask them to find all the 30's on their chart and identify how they could divide 30 with no remainders. Let them do the dividing using cubes if they need that additional support. Let them use the chart daily, gradually building up their memory of facts.

3 The children could pick two number cards from 0–9 and make a two-digit number that they recognise from the times tables, for example, 3 and 5 making 35. They should try to remember the table it is in, then check it on their chart. *"Which numbers will divide into 35 without leaving a remainder?"* (7 and 5)

Teacher-independent activities

Choose from:

1 In pairs, the children can match tables cards to answers cards.

2 Play the 'Share the biscuits' game (Resource sheet 44) again, restricting the children to the answers to the multiplications you are concentrating on if you prefer.

3 Using Resource sheet 1, make sure the children are seeing the patterns that the multiplication tables make on a 100 square, counting in 2's, 3's and so on.

Plenary session

- *"Is this true? 75 divides between 2 equally with no remainder? How do you know?"*
- *"If I want 5 children to have an equal number of biscuits each, what would be a good number of biscuits to have?"*

Fractions and decimals

Overall learning objectives

- Recognise simple fractions that are several parts of the whole.
- Recognise simple equivalent fractions.
- Order fractions on a number line, ordering amounts of money or measures.

Key words

one whole	tenth
part	decimal fraction
equal parts	decimal point
eighth	

Pizzas and cakes

Assessment focus

- Can the children recognise simple fractions that are several parts of the whole?

Resources

- Resource sheets 32, 45, 46 and 47
- paper strips, rectangles and circles, plastic shapes
- apples and other fruit, small cakes, chocolate bars and so on

With the whole class

- Do some paper folding with rectangles. Similarly, with a large circle fold and name fractions. Always make the point that it is a quarter or eighth of a *whole one*. Prepare a long strip folded into tenths. Leave these on display.

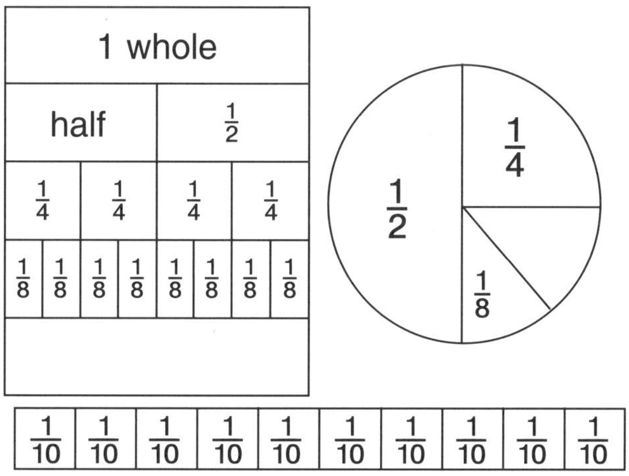

- Ask questions such as *"Sadia, come and point to an eighth"* and *"Which is larger, a whole one or a half?"*
- Move on to equivalent fractions using the folded paper. For example, ask *"What else could we call these two quarters here?"*

With the lower-achievers

With adult support

Choose from:

1 The children should cut along the solid lines only of the pizzas and cakes on Resource sheets 46 and 47 (not the dotted lines – those are there to show equivalent fractions). Help them to name the parts, then put them back together again to make whole pizzas and cakes. Stress that there are 10 whole ones.

2 The children can cut up the fraction cards on Resource sheet 45 and play 'Pizzas and cakes'.

Rules

i Shuffle the cards (take out the mixed numbers if you want) and put them face down.

ii The children take it in turns to take a card and identify a piece of cake or pizza (from Resource sheets 46 and 47) that is that fraction.

iii The fraction can be returned to the middle. Or play where the child keeps the fraction if they are right and play until everyone has had three goes. Observe – who names fractions confidently?

3 Cut up play dough or real apples and cakes so that children are quite clear about the use of fractions in daily life. A selection of chocolate bars shows parts of the whole in different ways.

Teacher-independent activities

Choose from:

1 The children can use paper to fold their rectangles and circles into fractions and then name the parts.

2 They can draw around some plastic shapes, cut them out and colour halves, quarters and so on.

3 On squared paper the children should make 4 x 4 squares and find as many ways as they can of colouring in half of the square. Extend this to colouring in just a quarter of each square.

Extend this again to using different rectangles, such as 5 x 2, and colouring in one tenth. Resource sheet 38 can be used to make a display of this.

Plenary session

- *"Which is larger, a half or a quarter?"*
- *"What about an eighth and a tenth?"*
- *"What have your learned today about fractions?"*

Which is larger?

Assessment focus

- Can the children recognise simple equivalent fractions and order them on a number line?

Resources

- a washing line and pegs
- larger versions of the fraction cards on Resource sheet 45
- three large pieces of paper the same size
- scissors and crayons
- folded paper from the 'Pizzas and cakes' activity above
- Resource sheets 29, 46 and 47
- a number line
- some dice

With the whole class

- Make the point that a quarter of an enormous pizza would give you much more to eat than a quarter of a tiny pizza.
- Remind the children *"When we compare and order fractions we are always talking about fractions of one whole."* Cut out the fraction cards on Resource sheet 45 and hang them on the washing line in order of size. With folded paper (a circle if you want to use pizzas again) cut out a half, a quarter, three quarters and so on. Compare them and order them. Then hang each one up with its matching fraction card. Extend this to eighths, tenths and so on, remembering to cut and fold from the same size paper each time. Leave the washing line on display for work with the lower-achievers.
- Focus on equivalent fractions of $\frac{2}{4}$, $\frac{1}{2}$, $\frac{4}{8}$ and so on, making more pizzas as you go along.

With the lower-achievers

With adult support

Choose from:

1 This is a complex aspect of maths for many children so you will need to continue with folding, cutting, ordering and naming fractions. Use the cut-outs from Resource sheets 46 and 47, but using only those circles that are the same size. Explain to the children *"When we are finding fractions, remember that they are always a part of a whole one."* Then ask them to find specific fractions. For example, *"Jack, find us a whole one." "Kas, find us a half of that whole one."*

2 Find and talk about equivalent fractions. *"Which is larger, a half or two quarters?" "Which fraction is the same size as five tenths?"*

3 Using Resource sheet 29 filled in as shown below, and with copies of the pizzas and cakes all the same size, the children throw a dice and move that many spaces around the track, picking up the fraction of cake or pizza that they land on. Play until one child has a complete pizza or cake, or so that the player with the most is the winner. Ask *"How much have you altogether now, Charlie? Do you have more than Cathy?"* and *"Look, this 2/10 is the same size as 1/5."*

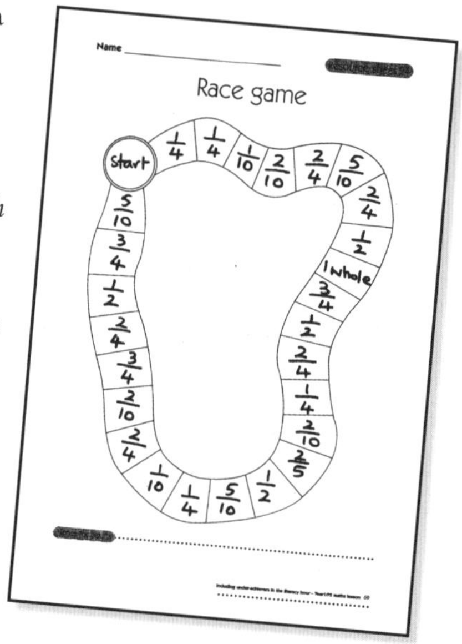

Teacher-independent activities

Choose from:

1 Let the children take the fraction cards off the washing line and put them in the correct order again on their own.

2 Give them other paper folding challenges, each child starting with the same size paper. They should fold, then colour in fractions, such as $\frac{2}{8}$ and $\frac{1}{4}$. You might want to demonstrate how to fold a circle into thirds so that you will get sixths, ninths and twelfths, as this gives wider experience in naming fractions. The children can then cut out the fractions made and put them in order on the washing line.

3 Give the children prepared sheets to complete such as the one below.

1.	$\frac{1}{4}$		$\frac{3}{4}$	1		
2.	$\frac{1}{3}$		1	$1\frac{1}{3}$	$1\frac{2}{3}$	
3.	$\frac{2}{5}$		$\frac{4}{5}$	1		$1\frac{2}{5}$
4.	$\frac{1}{10}$		$\frac{3}{10}$		$\frac{5}{10}$	

"What are the missing fractions?"

Plenary session

- *"If I divided a cake into 22/100/1000 pieces, what could we call each piece?"*
- Divide the class into two teams. Put the fraction cards from Resource sheet 45 face down and ask a child from each team to come and take one. The team with the smaller fraction wins half a point. Continue playing like this asking the lower-achievers to pin up the fractions in order on the washing line. If the two fraction cards are equivalent, the first team to point that out wins five whole points.
- Count along the number line in quarters – $\frac{1}{4}$, $\frac{1}{2}$, $\frac{3}{4}$, 1 whole, $1\frac{1}{4}$ and so on.

Fractions of numbers

Assessment focus

- Can the children find fractions of numbers including tenths and fifths?

Resources

- Resource sheets 42, 48, 49 and 50
- a large 100 square
- cubes, number lines and calculators

With the whole class

- Go over mental halving of numbers, linking this to the doubling that they know by heart. Demonstrate how easy it is to find a quarter if you find half of a half. You could do this with 12 or 24 (as at the bottom of Resource sheet 48).
- Find tenths of numbers, first physically with cubes, then using a large 100 square. Demonstrate that to find a fifth, you can add together two tenths, so 10 is a tenth of 100, and a fifth of a 100 is 20.
- Use Resource sheet 49. Cut it into four to play bingo. The children put numbers copied from the board randomly on their card. (Repeats are allowed.) You might want to focus on tenths of multiples of 10, so the child needs to write numbers between 1 and 10 on their card. You call out *"One tenth of 50"* and all 5's are covered with cubes. The first to cover a complete row is the winner.

With the lower-achievers

With adult support

Choose from:

1 Individually, ask the children closed questions, for example *"What is a third of 9?"* to assess who is grasping the concept.

2 Again individually, ask the children *"How can you find a quarter of a number once you know half the number? How can you find a fifth of 50?"*

3 Extend the activity to decimals as the children become more confident. Doing this with calculators can be successful, as this links fractions and decimals. For example, key in 35 divided by 10 and link this to dividing 35 cubes into tenths.

Teacher-independent activities

Choose from:

1 Resource sheet 48 is just for finding halves and quarters. Help the children to make the link to the halving that they can already do mentally. The lower part of the page with 12, 24, 48 and 96 is open-ended so that they can explore halving halves to find quarters. It might be useful to use calculators for 96 and go on dividing it by 2, so making the link with halving and dividing.

2 Resource sheet 42 can be used for the children to find halves of numbers (questions 1 and 2), thirds (questions 3 and 4), quarters (questions 5, 6 and 7) and fifths (questions 8 and 9). The quarter questions are half way down the page so that it is possible to cut the sheet in half, then copy and repeat the top part to make an easier sheet for finding just halves, thirds and quarters.

3 Using cubes and/or number lines, set tasks such as finding a tenth of 20.

4 Give those children who are able to tackle it Resource sheet 50. It links tenths and fifths.

Plenary session

- *"Abdul, tell us how to find a tenth on a calculator. Which keys did you use?"*
- *"Mary, find a tenth of 65 for us. What did you do?"*
- *"Without a calculator, what is an easy way to find a fifth?"*
- *"I found a tenth of a number on the calculator. The answer was 6.3. What was my number?"*

Calculator decimals

Assessment focus

- Can the children link decimals and fractions?

Resources

- calculators
- number lines, washing lines and pegs
- fraction and decimal cards
- prepared cards, such as 35 ÷ 10 = , 27 ÷ 10 = , 30 ÷ 10 = (all to give answers below 4 for use on a 0–4 number line)
- Resource sheet 51

With the whole class

- Put up a number line or washing line and count along in tenths. Tell the children that during the lesson they will be changing these fractions to decimals so that by the end of the lesson they will be counting in steps of 0.1.

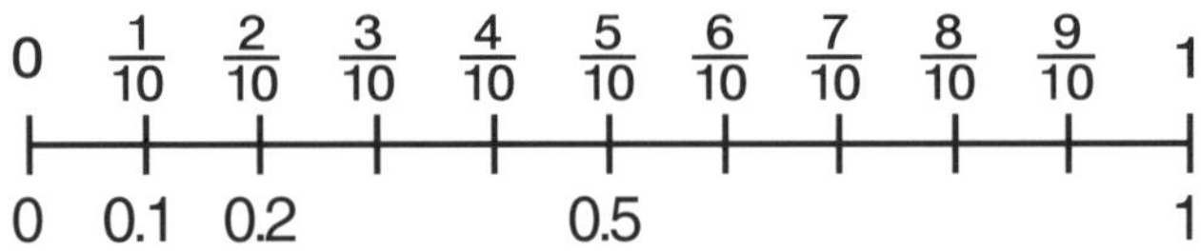

"Count along the line in tenths, then in decimals."

- In pairs, using calculators, the children enter the tenths fractions one at a time to see the pattern. For example, $\frac{3}{10}$ is entered as 3 ÷ 10. Explain that they should think of the line between the 3 and the 10 as the division sign. Ask some of them to write the decimals on cards and order them on the washing line. Count along the line, for example 0, 0.1, 0.2 and so on. Point out that $\frac{5}{10}$ is a half, which is the same as 0.5.
- Extend the activity by finding other decimals from fractions. *"How many ways can you make 0.5 on your calculator?"* (5 ÷ 10, 6 ÷ 12 and so on.)
- Repeat the activity, but this time start with a 0–2 number line.

With the lower-achievers

With adult support

Choose from:

1. With a number line go over the whole-class activity, making fractions into decimals and putting them on the number line. Gradually build up the number line to about five or six.
2. With a 0–1 number line, help the children to relate other fractions to decimals.

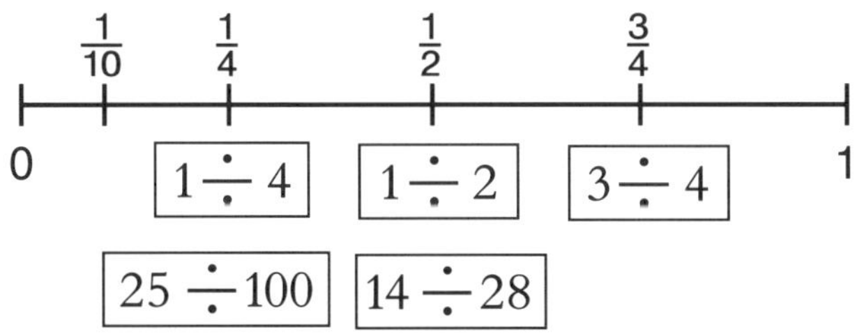

3. Check that all the children can count along a number line in steps of one tenth.

Teacher-independent activities

Choose from:

1. The children could work on Resource sheet 51 in pairs using a calculator. You might want to do the lower part of the sheet another day.
2. Give the children the prepared cards, for example '35 ÷ 10 =', and ask them to work them out and write the answers. They then put the cards in order, starting with the smallest answer, on a 0–4 number line.

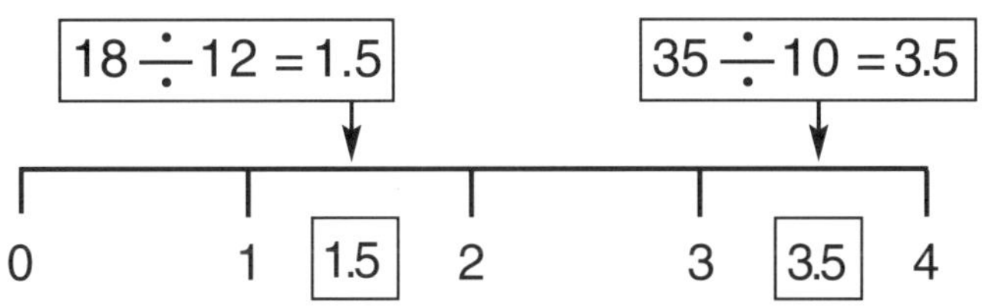

Plenary session

- *"What is a quarter of 100? What is three quarters of 100?"*
- *"Tell me a decimal that could come between one whole one and one and a half."*
- *"Who can tell us what they learned today about fractions and decimals?"*

Recognising more complex fractions

Assessment focus

- Can the children recognise fractions that are several parts of the whole?

Resources

- Resource sheets 45 and 47. Make the lower two circles into fractions, such as $\frac{1}{3}$ and $\frac{2}{6}$ by drawing in the lines for sixths, $\frac{8}{10}$ by drawing in the lines for tenths, and so on. Make various tenths, for example $\frac{6}{10}$, and $\frac{9}{10}$.
- several paper circles the same size as those on Resource sheet 47
- Resource sheet 52
- glue and paper
- scissors and crayons

With the whole class

- Again, using the folded paper from previous activities, remind the children of equivalent fractions that they have come across, then move on to find and name fractions such as $\frac{7}{10}$ and $\frac{2}{3}$.
- Ask the class questions such as *"If I divided a pizza into hundredths and gave you 3 of the bits, how much of the pizza would you have?"*

With the lower-achievers

With adult support

Choose from:

1. Give the group several copies of Resource sheets 47 and 52. They cut out shapes, outline in colour a fraction, such as $\frac{2}{5}$, name it and stick it on paper for discussion at the plenary session. Ask individuals to name fractions. Observe – who needs more help with this?
2. To support those who need more help, use Resource sheet 52 and help them to name each individual fraction, such as each of the fifths on the pentagon. Let them cut them out. *"Lisa, pick up two fifths. How many more fifths do you need to make a whole one? Now pick up enough fifths to make one whole one and two fifths."*

3 Make some more complex fraction cards (for example $\frac{3}{6}$, $\frac{3}{5}$, $\frac{2}{3}$ and $\frac{4}{5}$) to mix with those on Resource sheet 45. A pair of children turn over a card each. If a player can name their own fraction they win half a point.The player with the larger fraction also wins half a point.

Teacher-independent activities

This is a conceptually difficult area, so it needs as much initial adult support as possible.

Choose from:

1 As in the activity with adult support, ask the group to work on their own using several copies of Resource sheets 47 and 52. They cut out shapes, outline in colour a fraction, such as $\frac{2}{5}$, name it and stick it on paper for discussion at the plenary session.

2 Give each child Resource sheet 52 and some individual tasks to do on it. For example, *"Colour $\frac{1}{3}$ of the triangle red."* (This will need occasional adult supervision.)

Plenary session

- *"Yasmeen, pick up two fifths. Is she right?"*
- *"Is two fifths larger or smaller than a half?"*
- *"Do you think a third of a pizza is bigger or smaller than a quarter of a pizza?"*
- Ask the children to show the shapes they have cut out and coloured. They should tell the class the names of the shapes.
- *"Which of these shapes I'm holding up shows four sixths? three fifths?"* and so on.
- *"If you were very thirsty, would you rather have five tenths of a drink or three sevenths of it?"*
- Play 'Make a whole one.' Give each pair one minute to come up with an addition that totals one whole one.

Make a whole one

$$\frac{1}{10} + \frac{9}{10} = 1$$

$$\frac{1}{2} + \frac{1}{2} = 1$$

Ordering pounds and pence

Assessment focus

- Can the children order amounts of money?

Resources

- prepared cards with amounts of money written on them (75p, £1.50, £1.75 and so on) or cards of measures (2.1m, 3.2m and so on)
- an abacus
- Mega Money (from BEAM)
- coins
- Resource sheets 28, 29, 53, 71 and either 5 or 20

With the whole class

- Show the amounts of money on the cards and make sure the children can read them and can pick out the right coins (use Mega Money), they need to make up that amount. Point out the decimal point after the pounds. You might want to link these amounts of money to an abacus, demonstrating the decimal point.
- Repeat the activity another day, ordering cards of measuring amounts.

With the lower-achievers

With adult support

Choose from:

1 Using Resource sheet 53, give individual children different amounts of money to spend. Observe – who can cope with adding up amounts and with change? There are more items for sale on Resource sheet 71 so you can make your own shopping sheet with amounts to suit your children.

2 If the children need more help with money, use Resource sheets 28 and 29 to make money games (see the next page). Work with the children to teach them how to take the right amount of money, and encourage more able children to play the games at wet playtimes.

Play the game above using Resource sheet 28. Each child starts with 30p in their own pot. They throw the dice in turns and either add or subtract the amount they land on to or from their pot. The child with the most, or least, money at the end wins. Alternatively, play a game starting with £1 each in their own pot and subtracting amounts of about 10p to 15p that they land on. This requires an exchange of the pound when the first space is landed on.

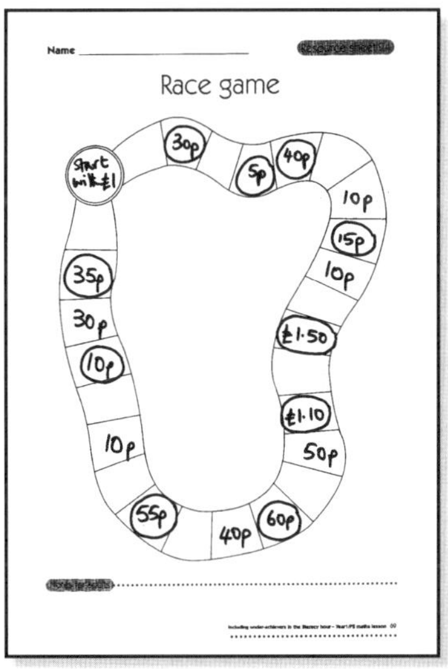

Play the game above using Resource sheet 29. Start with £1 each. Throw a dice to see how many spaces to move. The children win the amounts in the circles. Pay the bank amounts not in circles. Every time you pass 'Start' you win £1. For those still not sure of coins, devise an easier game on this sheet using coin stickers rather than written amounts. The children simply match the coins. Then you will need to help them to add up their totals at the end.

Teacher-independent activities

Choose from:

1 Give the children the prepared cards of amounts of money and ask them to put them in order starting with the smallest amount. They can then match the cards with the right amount of plastic money.

2 If the children are uncertain about the equivalence of amounts of money, ask them to make 25p/50p/£1 in five different ways, using Mega Money or other coins to demonstrate.

3 Use shopping catalogues and decide how to spend £20.

4 You could use Resource sheets 5 or 20 for ordering amounts of money written on the board.

Plenary session

- *"Who won the most money? Josh had £3.45 and Aklima had £3.54."*
- *"Which is more, £1.19 or £1.91?"*
- *"Sadia, come and write £3.56 on the board. Is she right?"*
- *"Sally paid for a box of sweets with four silver coins. How much could the sweets have cost? What change would she get from a five pound note?"*
- *"Alex bought a book with three silver coins all the same. What might the book have cost?"*
- *"What would you like to learn next using money?"*

Story maths and problems

Overall learning objectives

- Solve problems by choosing appropriate number operations.
- Use decimal notation and place value for tenths and hundredths in the context of measures and money. (See also Chapter 1.)
- Reason about numbers.

Key words

sign	equation
operation	number sentence
symbol	ten times larger

Telling stories

Assessment focus

- Can the children tell number stories and identify the operations used?

Resources

- cards with the operations (+, –, x and ÷) on them
- Resource sheet 54 and 58

With the whole class

- Tell some number stories, identifying the operation used for each one. For example, *"For the ugly bugs' ball, 5 drinks for each bug are needed. So if 10 ugly bugs are coming to the ball what is the number sentence we need?"* (Either 5 + 5 + 5 + 5 + 5 + 5 + 5 + 5 + 5 + 5 or 10 x 5). You could draw the four operations down the edge of the board as on Resource sheet 54 and ask some of the children to come to the front to map the story to the signs. Further stories around the theme of the ugly bugs' ball could be:

  "Each ugly bug needs 2 packets of crisps. How many are needed altogether?"

  "12 more ugly bugs come, so how many at the ball now?"

  "There are 24 buns to share between 12 ugly bugs. How many each?"

  "There are 30 cakes. How many of the bugs can have 5 each?"

  For each story, identify the operation used, show the card and ask the children to come to the front to write the number sentence.

With the lower-achievers

With adult support

Choose from:

1. Continue the starter activity. You will probably need to do this activity several times. Spend at least one session looking for key words that help to identify the operation used. 'Altogether' can often, but not always, mean addition. 'Leaves' is almost always used in subtractions.
2. When the children are ready, introduce money problems as well. (Some lower-achievers are very good at money problems.)

Teacher-independent activities

Choose from:

1. Ask the children to choose an operation card and, in pairs, make up a story with that operation.
2. Using Resource sheet 54, the children write the number sentences and match them to the operations around the edge of the sheet.
3. Resource sheet 58 gives further practice in identifying which operation is needed.

Plenary session

- Make the point that if you use multiplication to solve a problem, you could also solve it with addition and if you use division, you could also solve it with subtraction.
- *"When we need to solve a problem, we must think very carefully about which operation to use."*

Problems with money

Assessment focus

- Can the children work out money problems using multiplying and dividing by ten?

Resources

- an abacus
- place value/arrow cards
- coins
- Resource sheets 29, 55, 56, 57 and 58
- calculators
- counters and dice

With the whole class

- Demonstrate on an abacus how, when we multiply by 10, the digits move one place to the left. Don't say *"Add a nought"*! That is a trick and isn't true if you multiply a decimal number, such as 2.3 by 10. Emphasise the move to the left to multiply and the move to the right when we divide by 10.

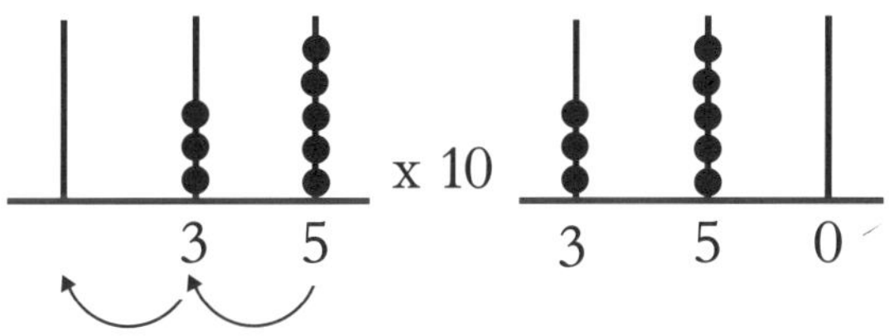

- Make sure that all the children can identify all the coins. If some can't, restrict yourself to one pences, ten pences and pounds. Make the point that a ten pence piece is 10 times as much as a one pence piece and a pound is 10 times as much as ten pence. Then work on simple problems, such as *"Crisps are 10 pence a packet. How much will 5 cost?"*, *"10 sweets cost 30 pence. How much does 1 cost?"* and *"10 drinks cost a pound. How much for 1?"*

With the lower-achievers

With adult support

Choose from:

1 Use coins, real if possible, to make up amounts of money, for example to make 80p. (Check that all the children can see that 10 one pence coins have the same value as 1 ten pence coin.) Then think about what would be 10 times smaller than 80p (8p) and 10 times larger (£8).

2 Using calculators, choose amounts of money to multiply by 10. Let the children explore this and move on to dividing by 10. Demonstrate how to write 320 pence as £3.20, talking about the decimal point, tenths and hundredths. Relate this to an abacus and to coins.

3 Use money as a context in which to explore the addition and subtraction of two-digit numbers, for example 23p add 45p and 65p, then spend 34p.

Teacher-independent activities

Choose from:

1 On Resource sheet 55, write some amounts of money in some of the spaces. You need a counter for each player, coins and a dice.

Rules

i Each player starts with £1. They take turns to throw the dice and move that many spaces.

ii If they land on a space with an amount of money, they win that much from the bank.

iii They then add up all their money and write down how much they now have.

iv The winner is the player who has the most money at the end – but only if they count it up correctly!

2 Using Resource sheet 56, each player starts with £1. This time the amount of money in the square must be paid to the bank and the amount of money in the circle is won.

3 You can make a game that covers the same learning but will look different by using Resource sheet 29.

4 Let the children use Resource sheet 57 when they are confident.

Plenary session

- Help the children to count up their money from the games.
- Go over multiplying and dividing by 10. *"What pattern can you see in the numbers when you multiply by 10?"*
- *"What did you learn today?"*

One-step problems

Assessment focus

- Can the children work out one-step problems?

Resources

- cards with number operations on them
- Resource sheets 59, 60 and 61

With the whole class

- Introduce some one-step problems. For example, *"When 10 more ugly bugs arrive at the ball there are 26 of them. How many of them were there to start with?"* (Identify this as a problem that needs subtraction.)

  "All the ugly bugs have 6 legs. How many legs altogether on 3 bugs?"

  "There are 18 packets of crisps left. If the grasshoppers eat 2 of them, how many will be left? Divide that number between 4 bugs. Is there a remainder? Which operations did you use?"

With the lower-achievers

With adult support

Choose from:

1 Repeat the whole-class activity with the group.

2 Set up a story context, such as a magician's magic garden where there are spiders (8 legs), bugs (6 legs), frogs (4 legs) and birds (2 legs). Tell the children that the magician wants the creatures to get into groups that have a total of 24 legs. So, 12 birds would have 24 legs between them, as would 3 spiders. Let the children work in pairs to make groups of 24 legs, at first just with one kind of creature in each set. (24 is a good number because it is 3 x 8, 6 x 4, 12 x 2 .) The children should choose an operation card for each calculation. Observe – who is confident with the operations? Record these on Resource sheet 59.

3 If some children need more help with operations, go back to telling them maths stories and selecting operation cards.

4 Introduce 'What's my number?' problems as on Resource sheet 60. Make sure all the children understand that both sides of the = sign must balance.

Teacher-independent activities

Choose from:

1 Resource sheet 60 involves very little reading. Let the children work in pairs to do this activity so that they have to talk about the calculations. (See also Chapter 4.)

2 Give them Resource sheet 61 which provides more one-step problems using money.

Plenary session

- Recap on balancing equations by doing some on the board, for example *"12 add something is 24. What is the something?"* Make sure all the children understand how to read equations using the words 'equals', or 'balances' or 'makes'. Clarify that = doesn't mean 'Now write the answer'!

Two-step problems

Assessment focus

- Can the children solve two-step problems?

Resources

- cubes
- a large class number line
- something to represent a bus
- operation cards
- Resource sheets 29, 62 and 63
- dice and counters

With the whole class

- Set up a story situation, such as the ugly bugs going to the ball on buses. Put some cubes on your 'bus' and count them. Say *"There are 18 bugs on the bus. 2 get off. How many are left on?"* Ask one child to hold up the right operation card. (Subtraction.)
- Move on to two-step problems. *"There are 18 bugs on the bus. 5 get off and 6 get on. How many bugs on the bus?"* Talk through what is happening, holding up the operation cards needed. Break the problem down into bits, for example *"18 subtract 5"*, then *"add on the 6"*. Ask for suggestions how to do it more quickly. (If 5 get off and 6 get on that is the same as saying there will be 1 more altogether.)

With the lower-achievers

With adult support

Choose from:

1 Go through the whole-class starter activity using different numbers. In pairs, the children could make up their own bus problems and record them in their own way. (This will give you insight into how they are thinking.)

2 Move on to problems that involve multiplication and division, such as *"There are 3 mini-buses with 12 bugs in each one and 1 mini-bus with 10. How many bugs altogether?"* and *"There are 12 grasshoppers, 10 bluebottles and 2 praying mantis who all want some cakes. There are 35 cakes. How many each?"* Talk about remainders.

3 Do some two-stage 'What's my number' activities as on Resource sheet 63. Make the point that brackets are done first.

Teacher-independent activities

Choose from:

1 Let the children do Resource sheet 62 in pairs.

2 Let them do Resource sheet 63 in pairs. (This would be a sensible place to use a calculator and then ask the children to tell everyone what they did at the plenary session.)

3 Play 'Make 24' with five 1–6 dice.

Rules

i The players take turns to throw all five dice. Then they can use as many of those dice as they want to make 24, using each number as often as they want. So with a throw of 6, 4, 2, 3 and 3, they could choose to ignore the 3's and make 24 with 6 + 4 = 10, which when multiplied by 2 is 20 and then 2 and another 2 is 24.

ii The others must check to see if they are right.

iii Players score 24 if they can make it, zero if they can't.

4 Play 'Target' on Resource sheet 29 filled in as shown below. Each player needs a counter and a 1–4 or 1–6 dice between them.

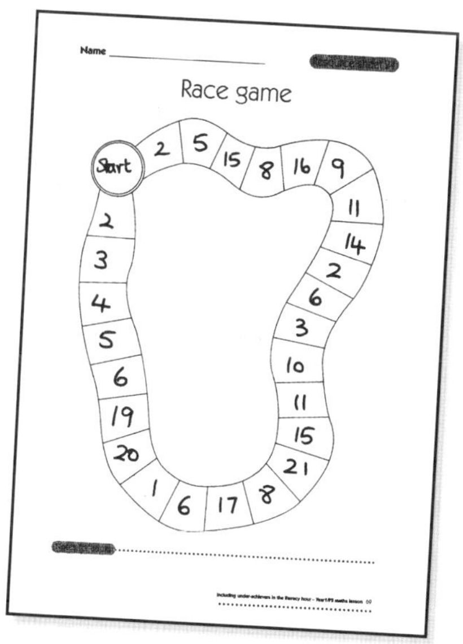

See over for the rules of the game.

Rules

i Choose a target number, for example 20.

ii The players throw the dice in turns and move that many spaces along the track. They write down the number they land on, for example 8.

iii For their next go, they add or subtract the number they land on. So, if they landed on 14 they would have 8 + 14 = 22. They must try to keep as close to the target number as they can. They should write all the sums down as they go.

iv If their score gets too high, they can choose to subtract the next number they land on. So, if they land on 6 they could say, 22 – 6 is 16.

v The winner has the score closest to the target number when they reach or go past the start.

Plenary session

- *"Often, to work out a problem we need to do it in several bits, sometimes using more than one operation."*
- *"Ask yourself 'What operation do I need to do this?'"*
- *"Read the question very carefully first. Look for those key words."*

Calculator games

Assessment focus

- Can the children solve simple problems on the calculator?

Resources

- calculators
- place value/arrow cards
- an abacus and some cubes
- Resource sheets 27, 30, 34, 64 and 65

With the whole class

- Remind the children how to partition numbers, for example 365 is 300 + 60 + 5. Ask them to partition a few numbers for you, such as 502.
- Play 'Space invaders' as in Chapter 1 (page 7). Extend the game to thousands and tens of thousands as the children become more confident, including zeros. For example, *"Use 2004 and talk about how to get rid of the zeros."*
- Take some of the numbers the children use in 'Space invaders' to divide or multiply by 10 and 100. For example, key in 340. Guess what the number will be if you divide by 10. Write it down, then divide and see if you were right.

With the lower-achievers

With adult support

Choose from:

1 Continue the whole-class starter games, observing carefully to see who is confident.

2 Introduce the children to Resource sheet 64 using an abacus if needed.

3 Move on to calculations like those on Resource sheet 65 where the operation needs to be worked out. For example, 34 ☐ 10 = 340. Encourage the children to think logically about this and also to do rough mental calculations. If an answer is quite a bit larger than the starting two numbers it is likely that the operation is multiplication (if the numbers are whole numbers) and so on.

Teacher-independent activities

Choose from:

1 Complete Resource sheet 65 using a calculator. Encourage the children to use the constant key. Often 10+ = = = = = will go on adding tens (or try 10 + + = = = = =). If that doesn't work, try a different make of calculator.

2 Use Resource sheet 34 to show how division undoes multiplication. (You can do this without using a calculator if you want, or explore larger numbers with a calculator.) White out the numbers of the example and substitute numbers, such as the example below.

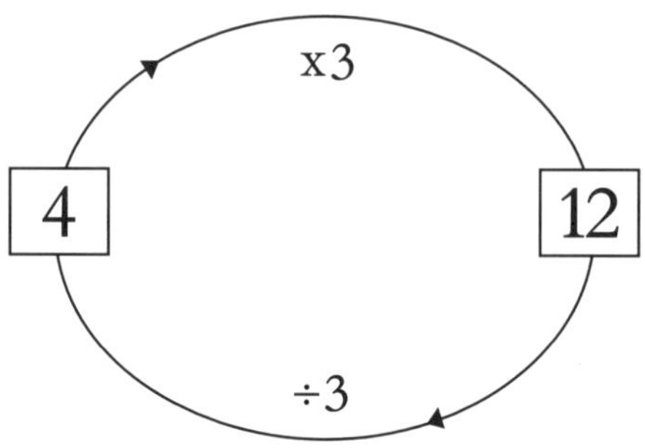

3 Use Resource sheet 30 again, this time with a much larger number, for example making pairs to 500. Let the children use calculators.

4 Use Resource sheet 27 again, using large numbers to practice multiplying and dividing by 10 and 100. This example could probably be done without a calculator, but it would also be an activity to introduce decimal points using a calculator. Each player needs their own colour of cubes.

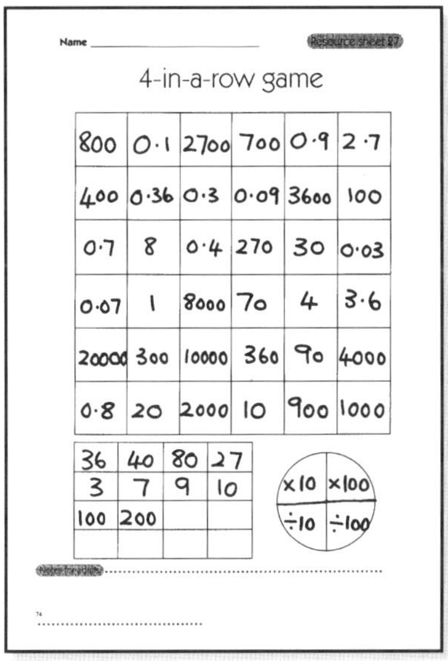
Name ……………………

4-in-a-row game

800	0·1	2700	700	0·9	2·7
400	0·36	0·3	0·09	3600	100
0·7	8	0·4	270	30	0·03
0·07	1	8000	70	4	3·6
20000	300	10000	360	90	4000
0·8	20	2000	10	900	1000

36	40	80	27
3	7	9	10
100	200		

Rules

i Take it in turns to choose a number from the big grid, for example 270. Decide which number you need from the little grid and which operation you need from the circle. For example, to make 270 you can have 27 x 10.

ii Cover the number on the big grid with your colour cube.

iii The winner has 4 of their numbers in a row.

Variation: It changes the game a bit if you make the circle into a spinner and you have to spin to find your operation. This makes a longer game.

Plenary session

- *"Tell us something you found out on your calculator."*
- *"Who can describe the pattern that you get when you divide by 10?"*
- *"What did you enjoy today?"*
- *"Remember that you are likely to press the wrong key sometimes, so always make a rough answer in your head when you use a calculator."*

Measuring problems

Assessment focus

- Can the children work out simple measuring problems?

Resources

- number lines
- measuring equipment
- Resource sheets 66 and 67
- calculators

With the whole class

- Remind the children about reading numbers on number lines. Make sure they can all write in missing numbers on unmarked lines. (See Resource sheets 11 and 12.)
- Give them experience in reading scales. There is no short cut to this! Children must have the practical experience.
- Build up the children's knowledge of measuring units again focusing on multiplying and dividing by 10, working out half a kilogram and half a kilometre and so on. Say *"Jessie is 150cm tall. How much is that in millimetres?"* Establish that you multiply by 10. *"1 kilometre is 1000 metres. How many metres in half a kilometre/10 kilometres?"*
- Give simple problems of addition and subtraction of measures, such as *"What is half a litre plus 200ml?"*

■ Use measuring as a context in which to work at mental addition and subtraction of two-digit numbers using a number line.

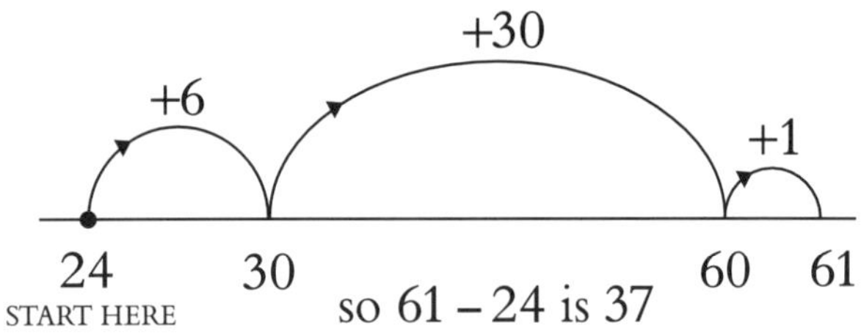

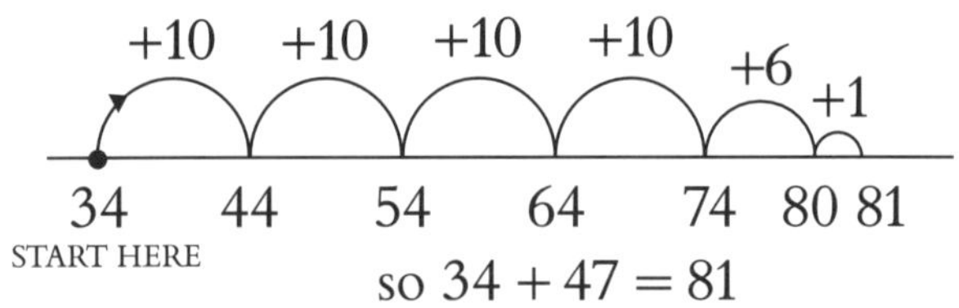

With the lower-achievers

With adult support

Choose from:

1 Set up some practical measuring situations where the children need to read a scale.

2 Using calculators, let the children explore larger numbers. Work on Resource sheet 66, helping the children to see the patterns in the numbers.

Teacher-independent activities

Choose from:

1 Let the children complete Resource sheet 67 in pairs.

2 Ask the children to weigh and measure books, parcels and so on using weights. Ask them to keep a record of these weights so that you can check them at the plenary session.

Plenary session

■ Show the children how to add up a set of weights, starting with the largest one, for example *"500 grams, plus 100, plus 20, plus 10".*

■ *"Is everyone sure that they can read all of these scales? Who would like more practice with that?"*

Rounding up or rounding down

Assessment focus

■ Can the children work out whether they need to round up or round down?

Resources

■ number lines

■ cubes

■ Resource sheet 68

With the whole class

■ Set up some story situations, such as a party or a trip to the seaside. Pose some questions that need either rounding up or rounding down. *"Each coach can take 50 people to the seaside. 234 people want to go. Let's count up in 50's on the number line."*

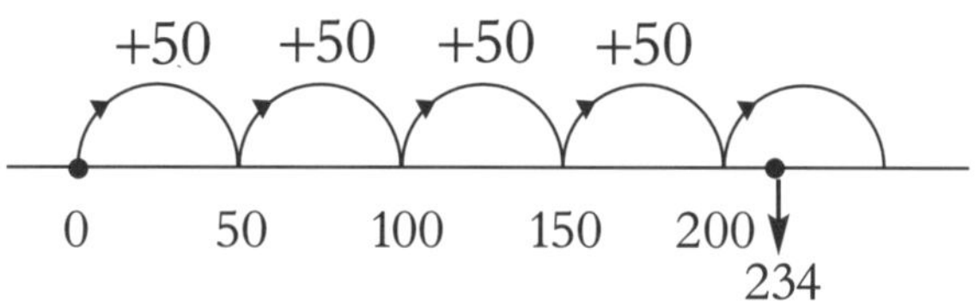

"One hop of 50 is for one coach, a second hop of 50 is for 2 coaches… and so on."

"We need 4 coaches to take 200 people and our next hop of 50 gets us to 250 people. In pairs, work out how many coaches would be needed." Establish that with this problem you would have to round up because you can't have a bit of a coach and all the people want to go on the trip, so you need 5 coaches.

■ *"There are 6 toffee apples to a box. How many boxes would I need to pack 17 of them?"* This is another rounding up problem because you need to pack all the toffee apples.

■ Tell a rounding down problem. For example, *"If I have 17 toffee apples, how many boxes can I fill?"* This rounds down to 2 boxes and 5 left over.

- Money problems are almost always rounding down problems because usually there is not enough money to buy the next item. For example, *"Tickets cost £5 each. Jane has £24. How many can she buy?"* She will have some money left over but not enough to buy the 5th ticket.

With the lower-achievers

With adult support

- Using a number line go through some more problems identifying with the children whether they need to round up or round down. For examples see Resource sheet 68.

Teacher-independent activities

Choose from:

1 Let the children complete Resource sheet 68 in pairs, ticking the round up or round down box each time.

2 Ask the children to make up some problems for others to identify whether it is rounding up or rounding down.

Plenary session

- This is an area of maths that some children find confusing, so they need to do plenty of problems until they can work out the logic of when to round up and when to round down.
- *"Lubna and Yasmeen, tell us your money story where you have to round down because you don't have enough money."*
- *"Explain to us why you have to round down with that problem."*
- *"Who feels sure they understand this now? Who would like more help?"*

Number squares – 1

1	2	3	4	5	6	7	8	9	10
11	12	13	14	15	16	17	18	19	20
21	22	23	24	25	26	27	28	29	30
31	32	33	34	35	36	37	38	39	40
41	42	43	44	45	46	47	48	49	50

1	2	3	4	5	6	7	8	9	10
11	12	13	14	15	16	17	18	19	20
21	22	23	24	25	26	27	28	29	30
31	32	33	34	35	36	37	38	39	40
41	42	43	44	45	46	47	48	49	50
51	52	53	54	55	56	57	58	59	60
61	62	63	64	65	66	67	68	69	70
71	72	73	74	75	76	77	78	79	80
81	82	83	84	85	86	87	88	89	90
91	92	93	94	95	96	97	98	99	100

Notes for adults

Name ______________________________

Number squares – 2

1	2		4	5		7	8	9	10
	12	13		15	16	17	18	19	20
21	22		24			27	28	29	30
31	32	33	34	35			38		40
41	42				46			49	50

1		3	4		6	7	8	9	
11				15	16			19	20
21	22	23	24	25	26	27		29	
31		33			36	37		39	40
41	42	43	44	45	46		48	49	
51	52	53	54		56	57	58	59	60
61	62	63	64	65	66		68	69	
71	72		74	75	76		78	79	80
							88	89	
91	92	93	94	95	96	97	98	99	100

Notes for adults

Name ______________________________

Resource sheet 3

Missing numbers

1

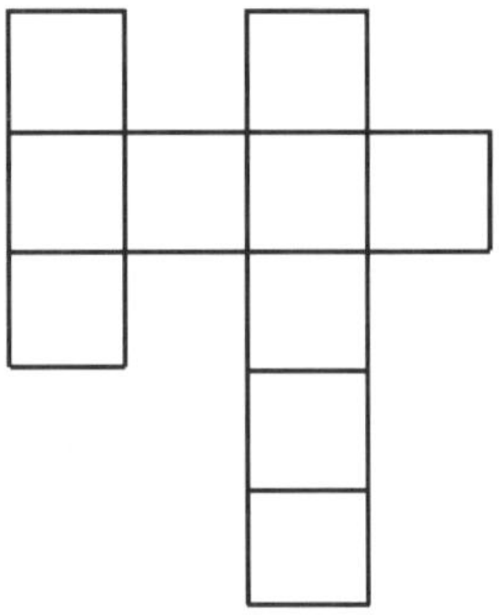

2

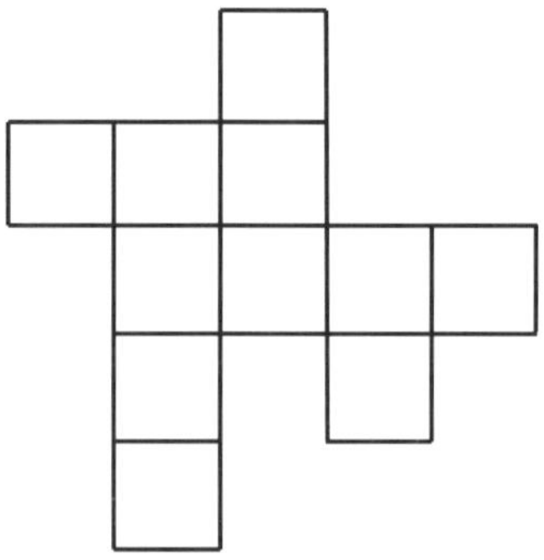

3

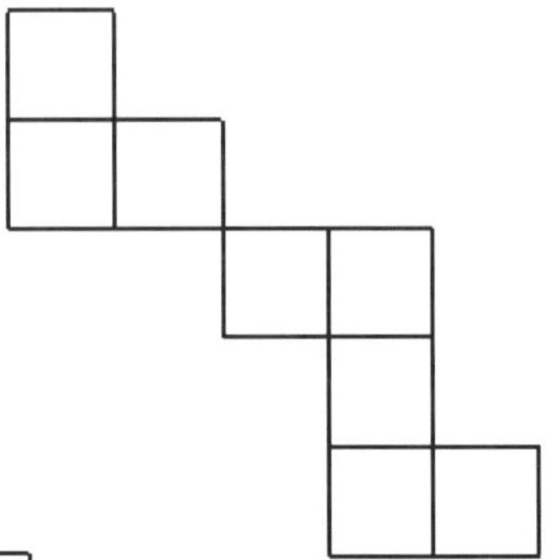

4

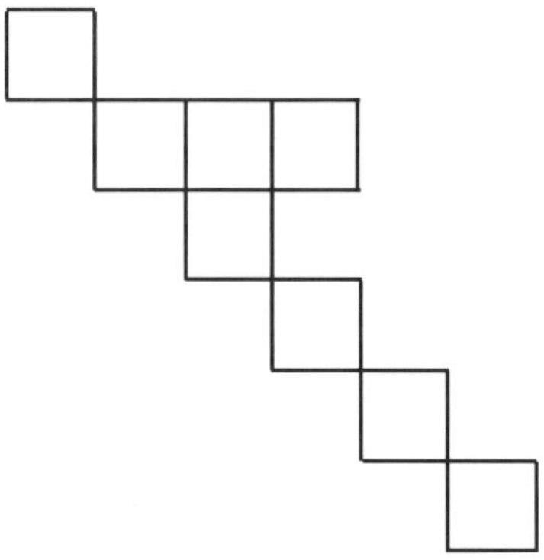

5

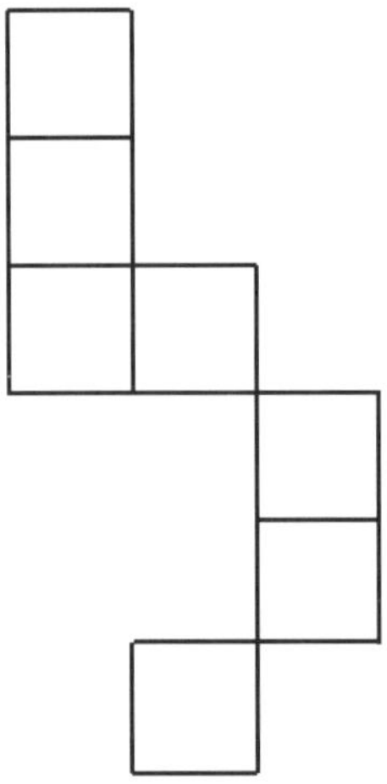

6

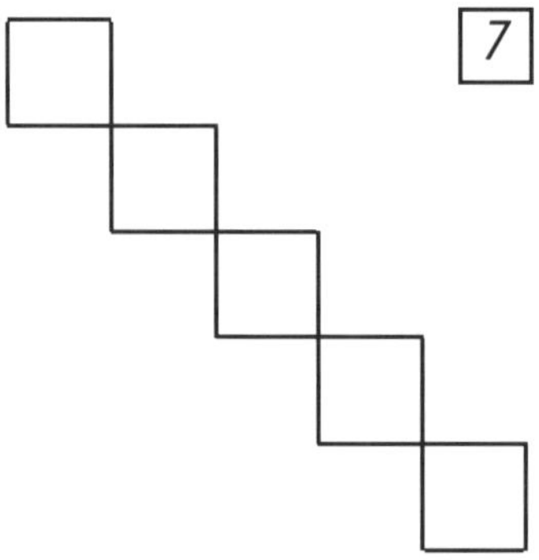

7

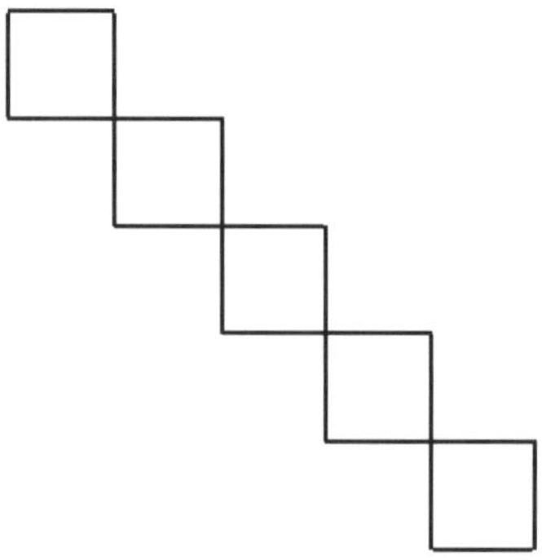

Notes for adults

Name ______________________________

Number squares – 3

Notes for adults

Name ________________________________

Washing lines

1.

2.

3.

4.

5.

Notes for adults

Name ______________________________

Resource sheet 6

Three dicey numbers

hundreds	tens	ones

I made

1.

2.

3.

4.

5.

Notes for adults

Name ______________________________

Resource sheet 7

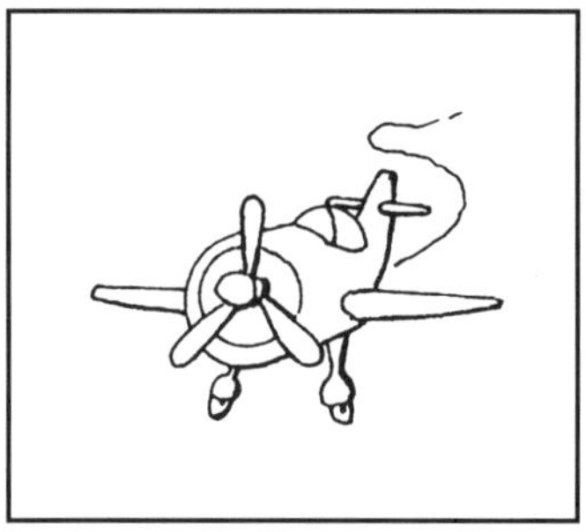

Abacus hundreds

1.

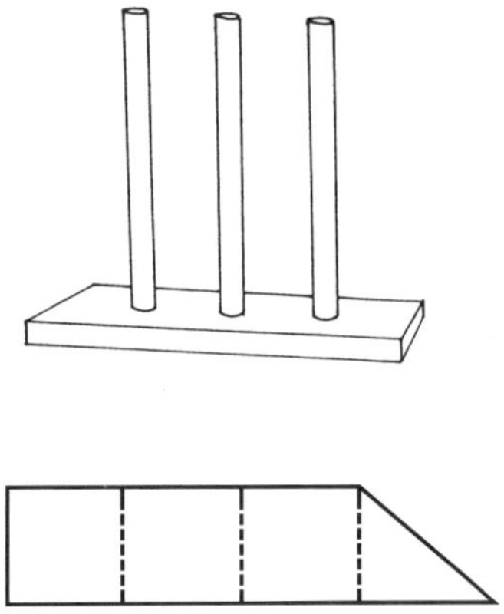

2.

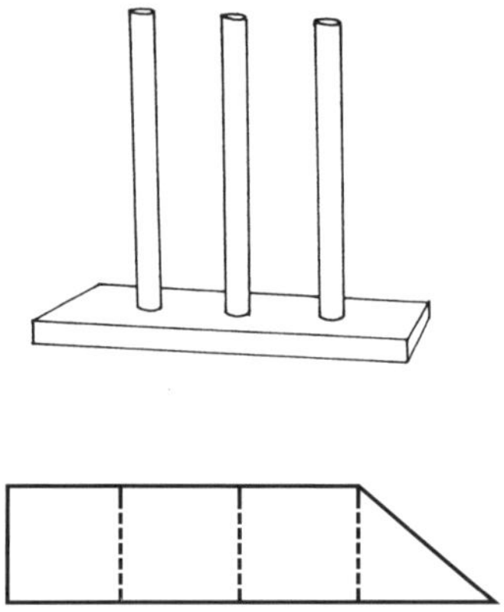

3.

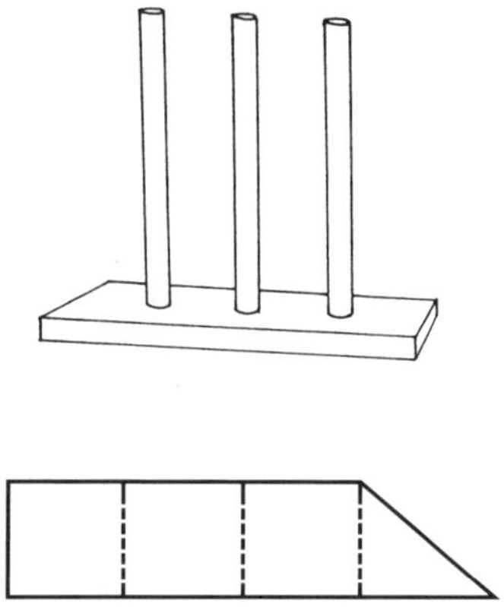

4.

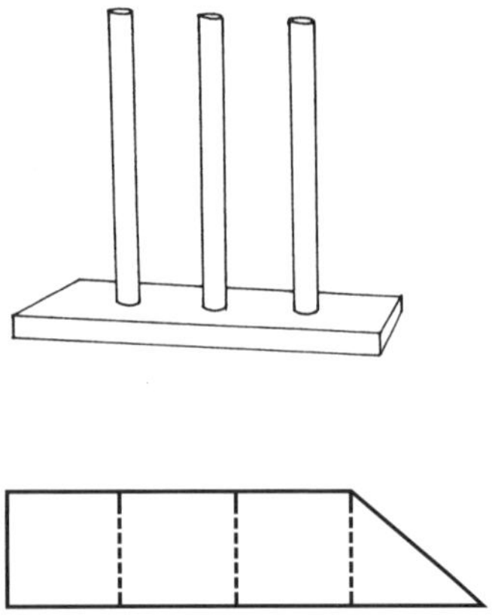

5.

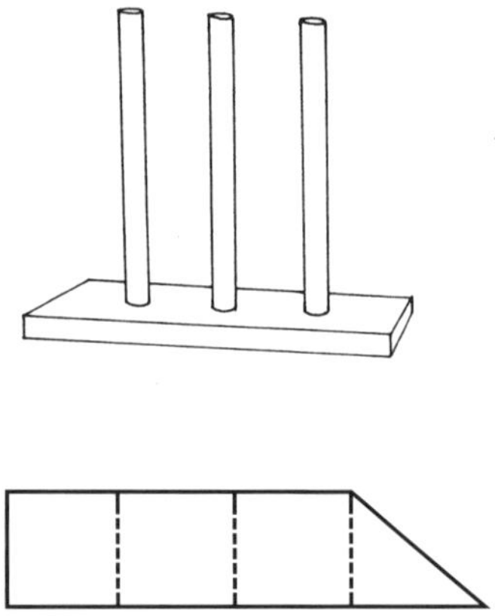

6.

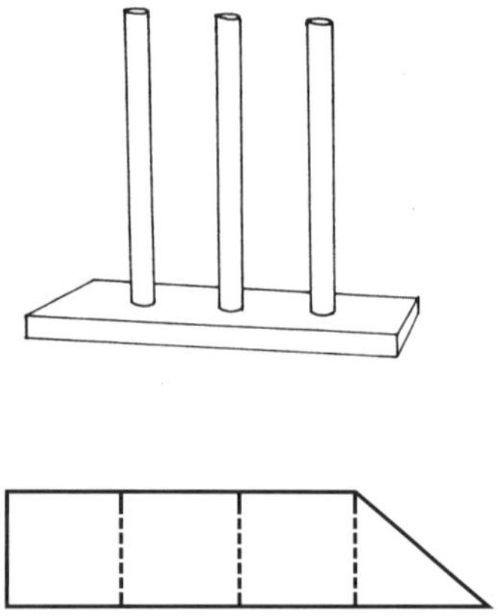

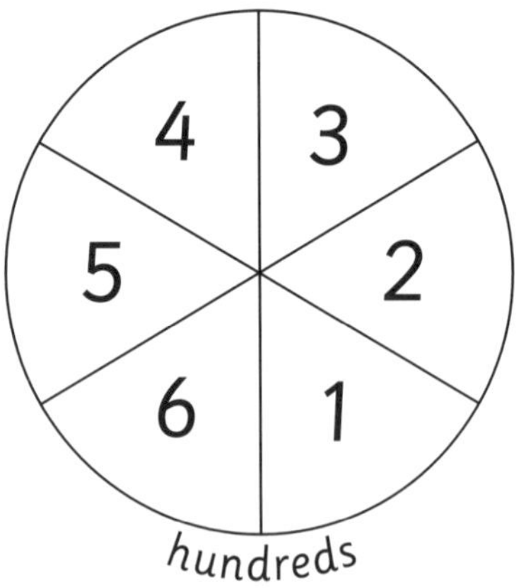

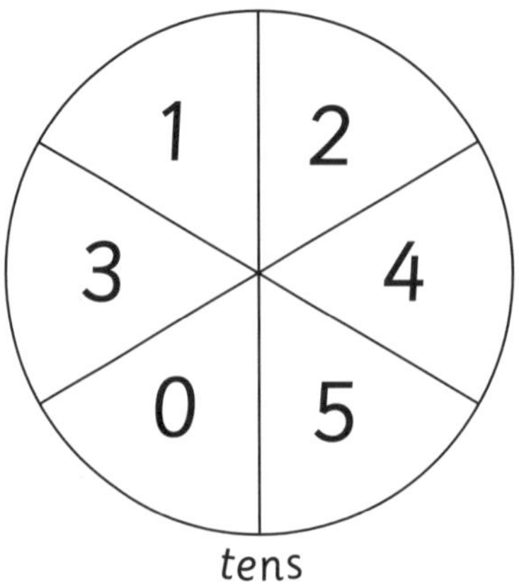

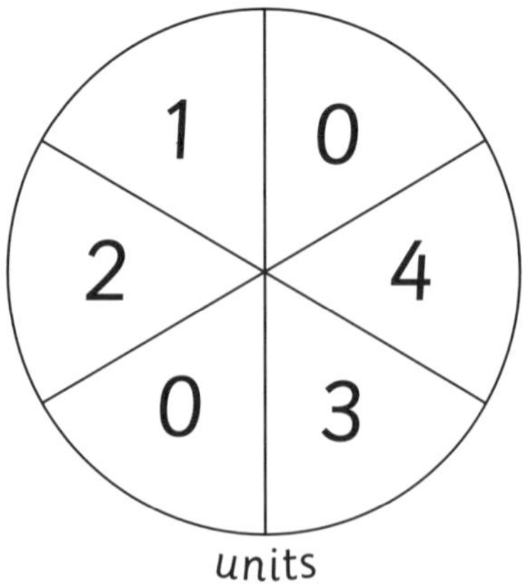

Notes for adults

The three circles above are the spinners and should be cut out and mounted on card. Use a paper-clip around the point of a pencil as the spinner.

Name ______________________________

Resource sheet 8

Greater than?

					Yes	No
		>				
		>				
		>				
		>				
		>				

				Yes	No

Notes for adults

Name ______________________________

Number track – 1

0

1 2 3 4 5 6 7 8 9

10

19 18 17 16 15 14 13 12 11

20

21 22 23 24 25 26 27 28 29

30

39 38 37 36 35 34 33 32 31

40

41 42 43 44 45 46 47 48 49

50

59 58 57 56 55 54 53 52 51

60

61 62 63 64 65 66 67 68 69

70

79 78 77 76 75 74 73 72 71

80

81 82 83 84 85 86 87 88 89

90

99 98 97 96 95 94 93 92 91

100

Notes for adults

Name ______________________________

Number track – 2

Notes for adults

This sheet fits exactly onto Resource sheet 9 in order to make a track beyond 100.

Number lines – 1

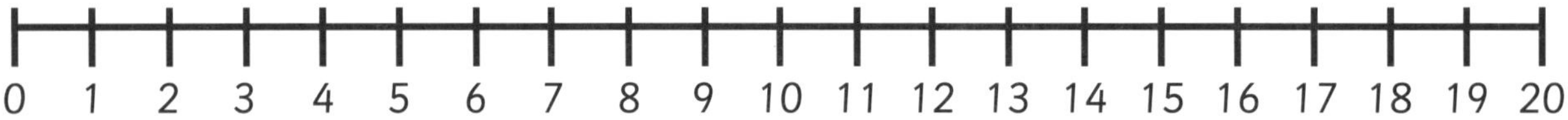

Notes for adults

This sheet can be cut up to make a long number line or used as a recording sheet for calculations. You can change the picture to one from Resource sheets 69 to 71.

Name ______________________________

Resource sheet 12

Number track – 1

1. Fill in the numbers.

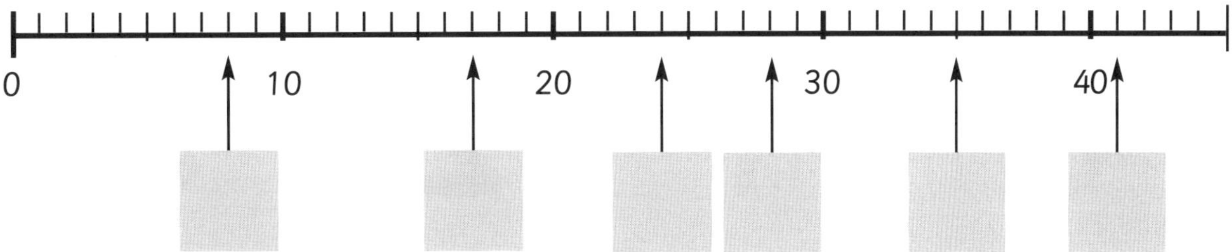

Round these to the nearest 10.

2. 8

3. 17

4. 24

5. 28

6. 35

7. 41

Fill in the missing numbers.

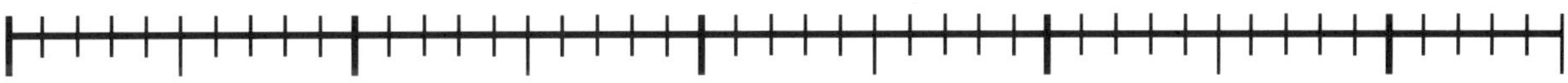

Round to the nearest

Notes for adults

Number cards

0	1	2	3
4	5	6	7
8	9	10	11
12	13	14	15

Number cards

16	17	18	19
20	21	22	23
24	25	26	27
28	29	30	31

Number cards

32	33	34	35
36	37	38	39
40	41	42	43
44	45	46	47

Number cards

48	49	50	51
52	53	54	55
56	57	58	59
60	61	62	63

Number cards

64	65	66	67
68	69	70	71
72	73	74	75
76	77	78	79

Number cards

80	81	82	83
84	85	86	87
88	89	90	91
92	93	94	95

Number cards

96	97	98	99

Name ______________________________

Resource sheet 20

Number sequences

1.

2.

3.

4.

5. The rule is ________

6. The rule is ________

Notes for adults

Name ______________________________

Odd and even

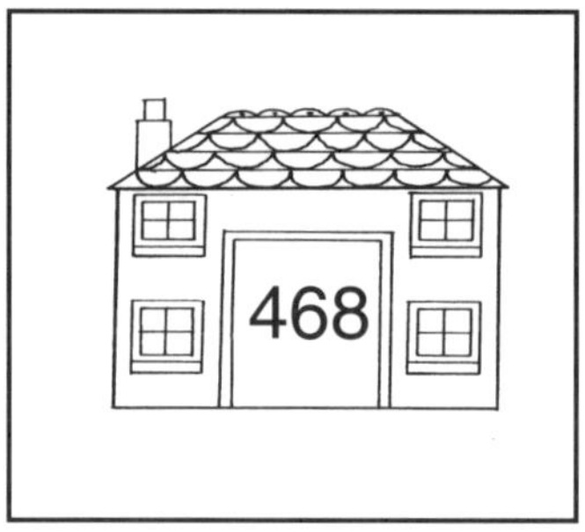

1. Odd numbers end in...

2. Even numbers end in...

Write T for true or F for false.

2	is an odd number.	F
3. 3	is an even number.	
4. 15	is an odd number.	
5. Odd number + odd is even.		
6. Odd + even is odd.		
7. Even + even is even.		
8. Odd + odd + odd is even.		

Notes for adults

Name ______________________________

Number sequences – 1

	Multiples of	Not multiples of
Odd numbers		
Even numbers		

Notes for adults ....

Name ______________________________

Resource sheet 23

Number lines – 2

2+7+3

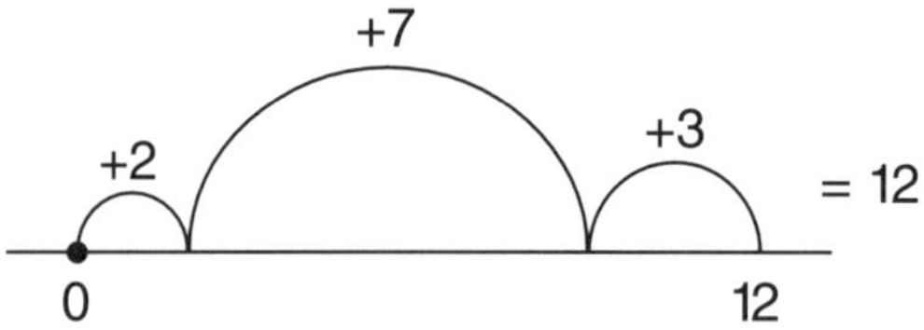

1. ______________________________ =

2. ______________________________ =

3. ______________________________ =

4. ______________________________ =

5. ______________________________ =

6. ______________________________ =

Notes for adults

Name ______________________________

Split it up

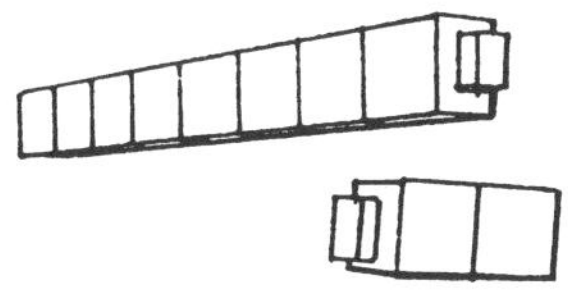

8 + 2 = 10

Start with ☐ cubes.

Notes for adults ....

....

Number sentences

and	add	plus	+	
altogether makes	makes			
more	is more than	is		
equals	+	+	=	=
take away	minus	subtract		
the difference between	and	is		
leaves	–	–	=	=

Notes for adults ....

....

Name ______________________________

Resource sheet 26

3 numbers game

Use these 3 numbers.

[ ] [ ] and [ ]

1. [ ] add [ ] makes [ ]

2. [ ] plus [ ] equals [ ]

3. [ ] and [ ] altogether makes [ ]

4. [ ] subtract [ ] leaves [ ]

5. [ ] take away [ ] is [ ]

6. [ ] count back [ ] equals [ ]

Notes for adults ....

....

Name ______________________________

4-in-a-row game

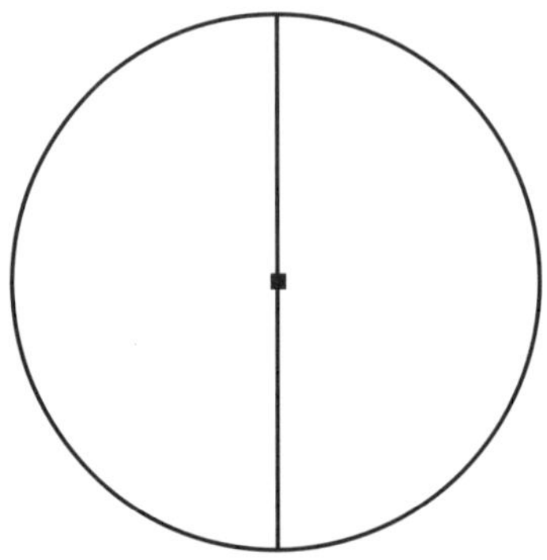

Notes for adults

Name ______________________________

What's in the pot?

Notes for adults

Name ______________________________

Race game

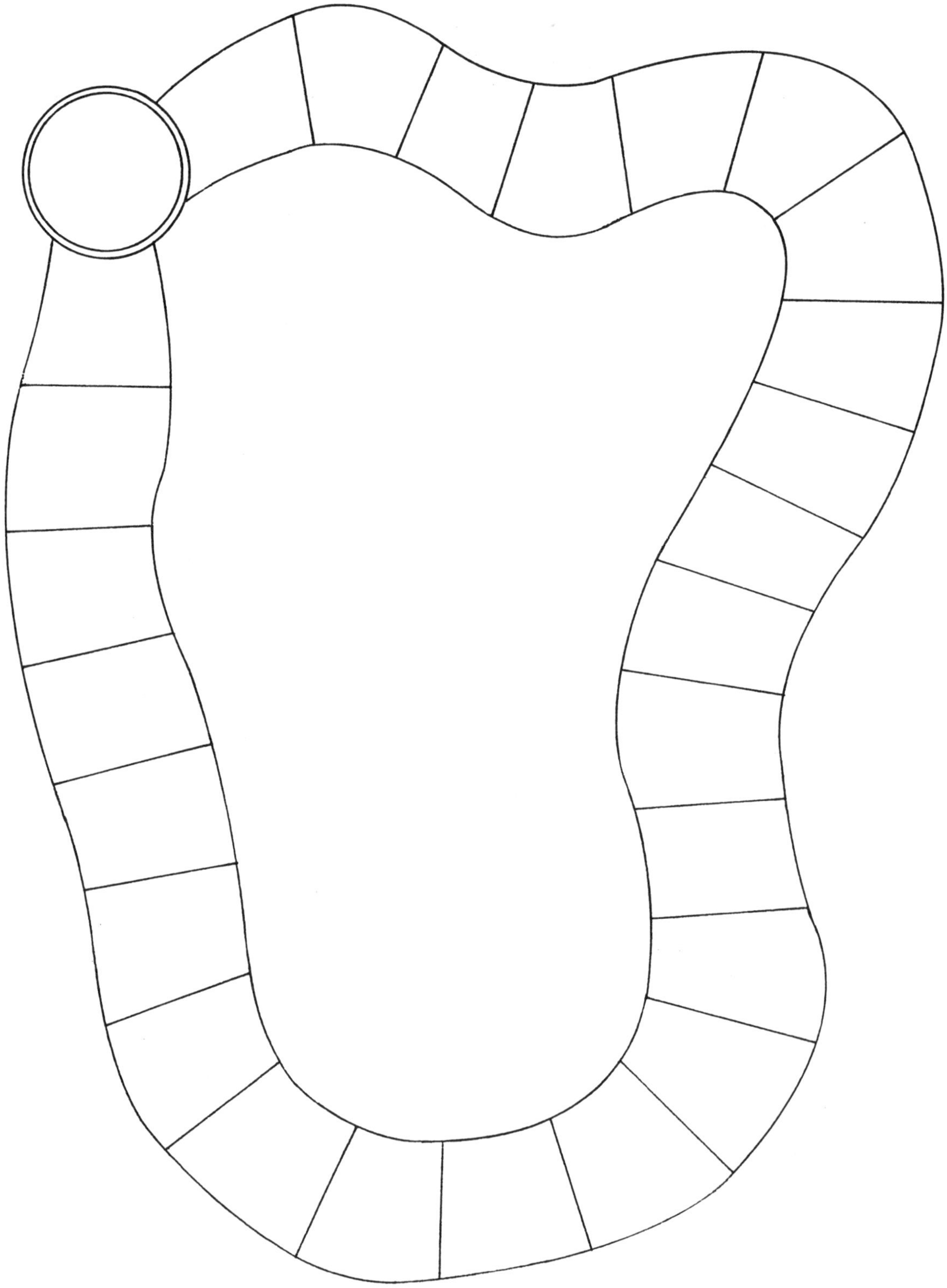

Notes for adults ....

....

Name ______________________________

Dinosaurs

Make pairs to [ ]

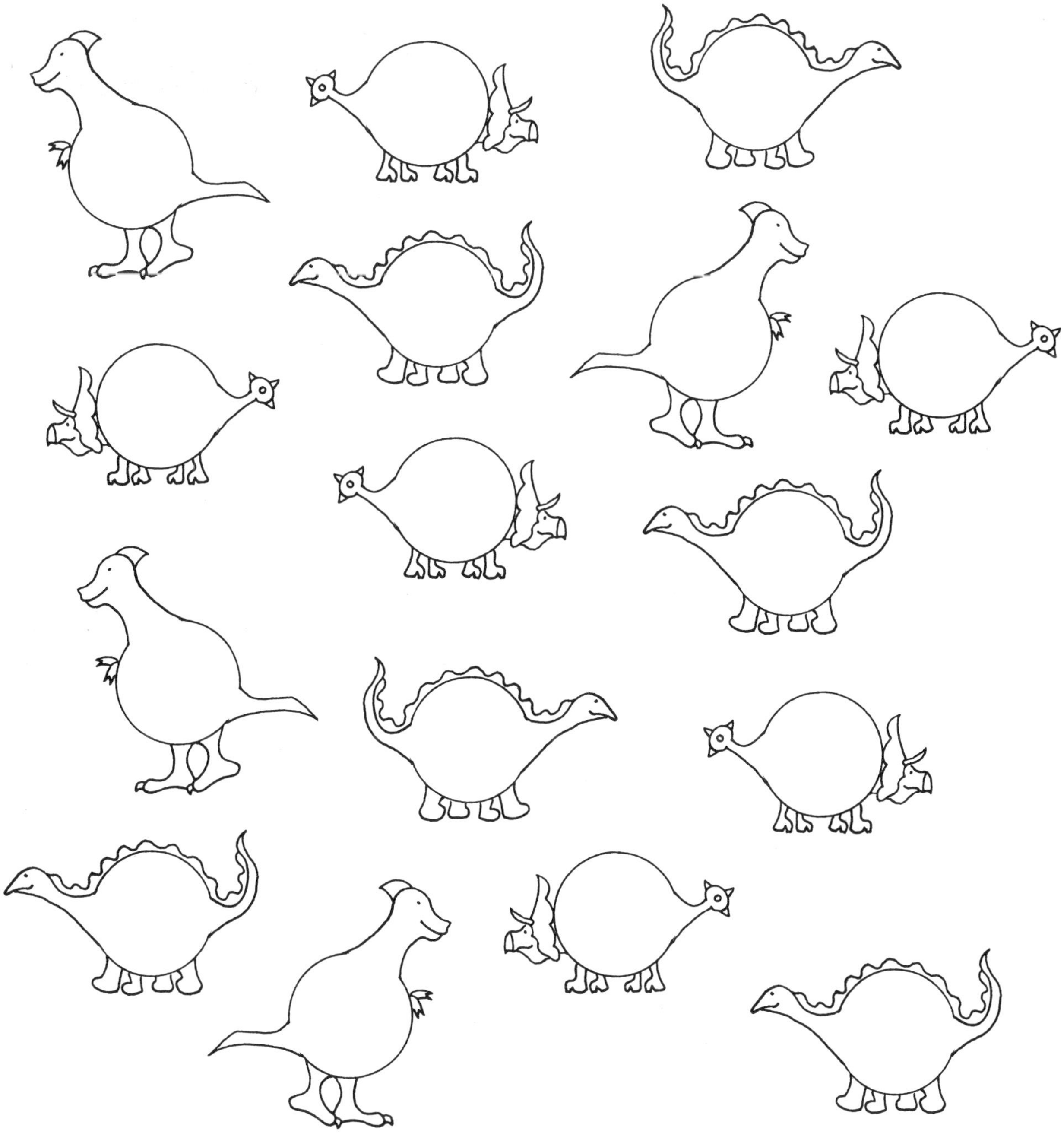

Notes for adults ....

....

Name ______________________________

Find easy pairs

Bill

7p
4p
3p
6p

7p + 3p → 10p
4p + 6p → 10p

[20] p

1\. 8p
6p
2p
4p

8p + 2p → 10p

[ ] p

2\. 7p
6p
3p

[ ] p

3\. 18p
7p
2p
3p

[ ] p

4\. 27p
16p
4p
3p

[ ] p

5\. 13p
6p
17p
4p

[ ] p

6\. 24p
15p
5p
2p
6p

[ ] p

Cake 13p

Lolly 17p

Juice 24p

Notes for adults ....

....

Name ______________________________

Clowns

Use ____ clowns.

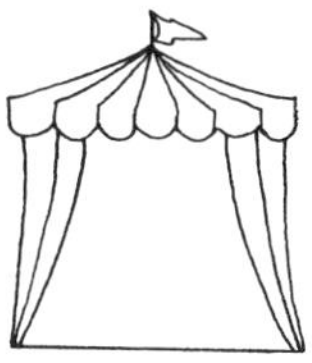 in ____ out

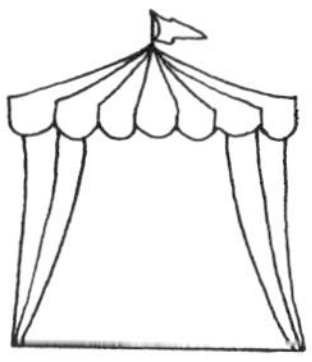 in ____ out

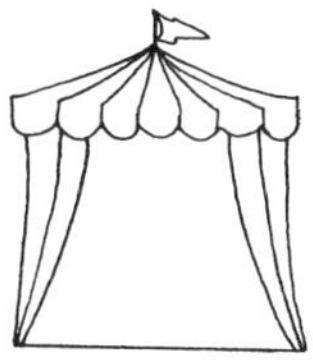 in ____ out

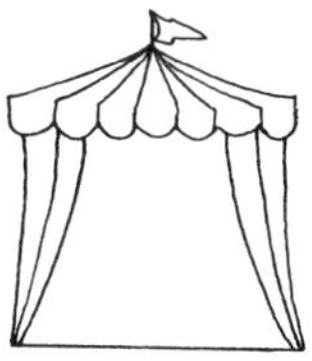 in ____ out

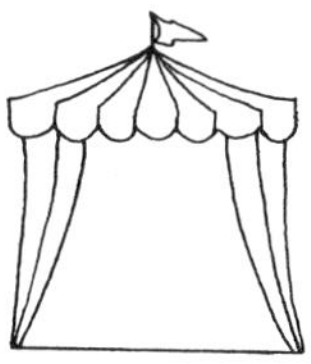 in ____ out

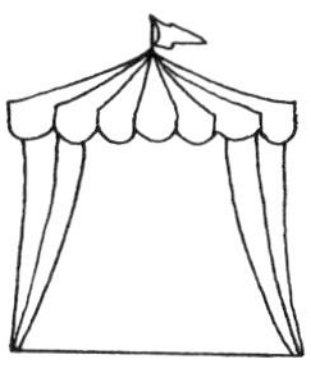 in ____ out

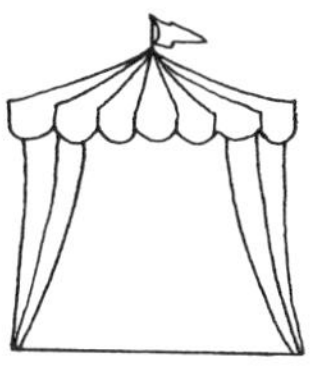 in ____ out

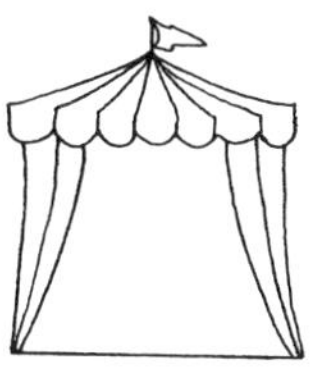 in ____ out

Notes for adults

Name ______________________________

If I know...

If I know that 7 + 3 = 10

I also know that 3 + 7 = 10

and 10 − 3 = 7

and 10 − 7 = 3

1. If I know

   I also know

   and

   and

2. If I know

   I also know

   and

   and

3. If I know

   I also know

   and

   and

4. If I know

   I also know

   and

   and

Notes for adults

Name ______________________________

Resource sheet 34

Step forward, step back

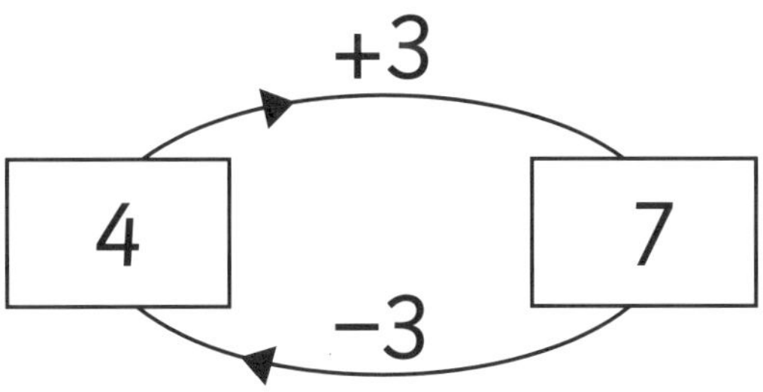

1.

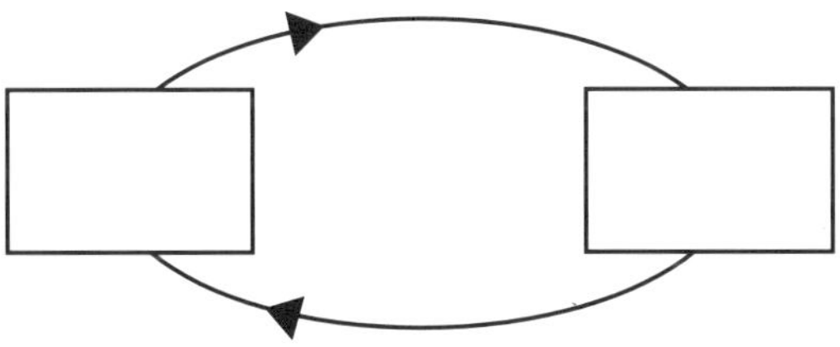

2.

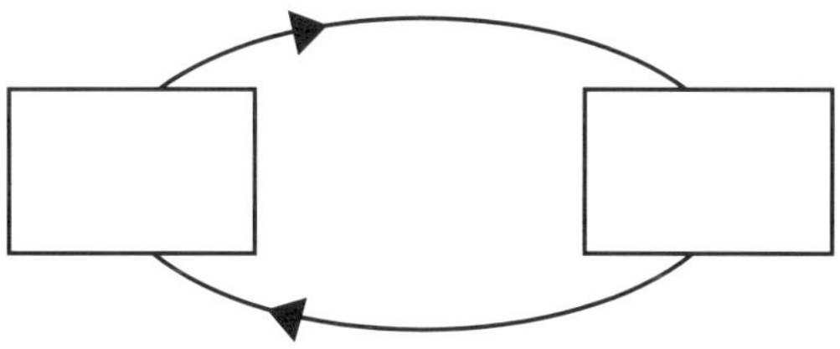

3.

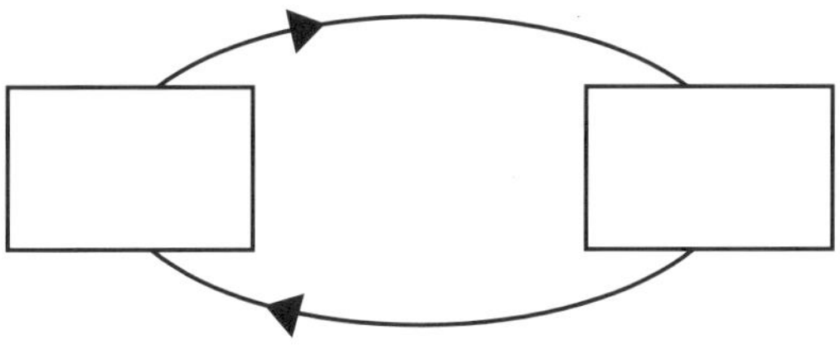

4.

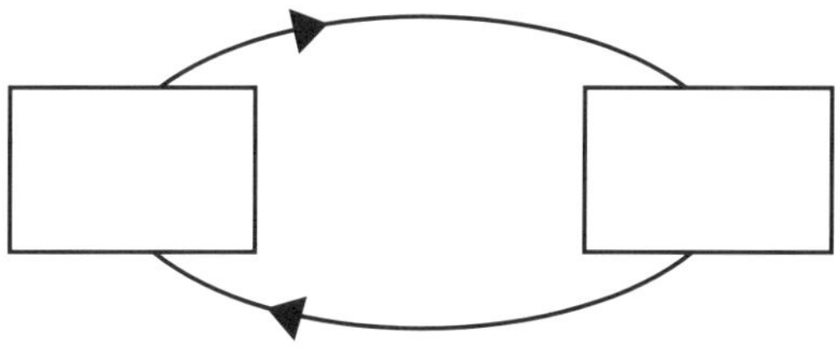

5.

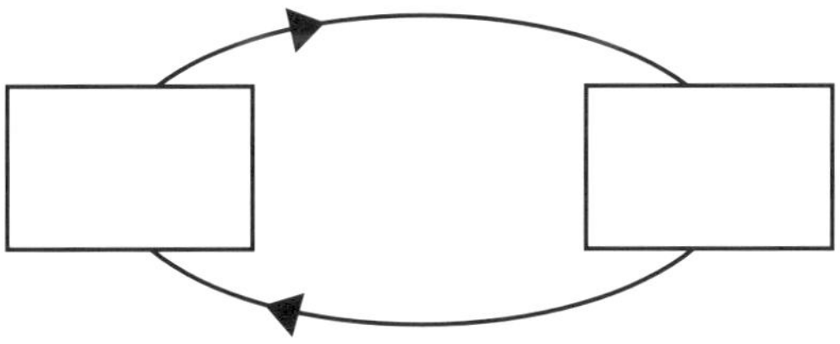

6.

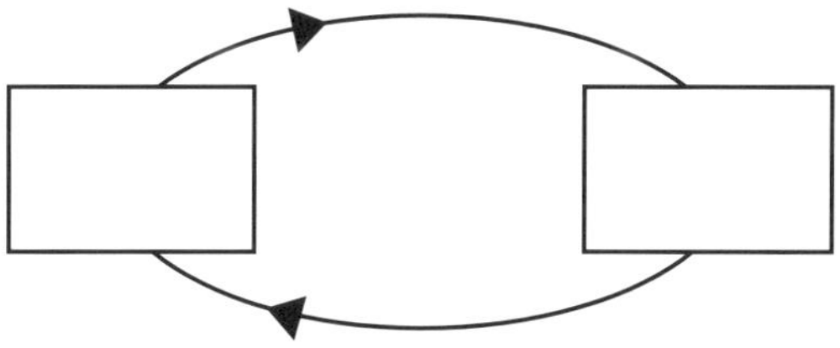

7.

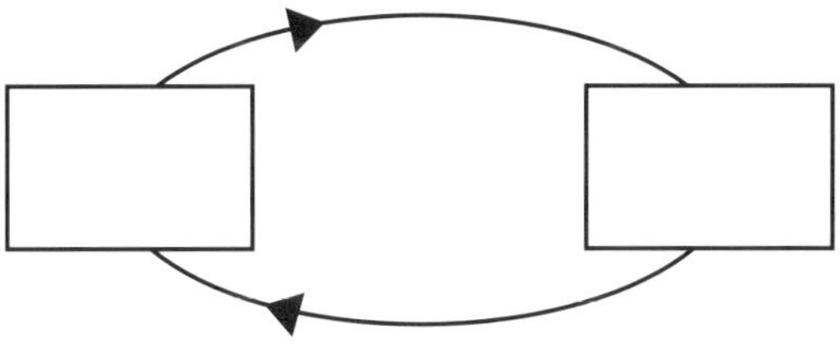

8.

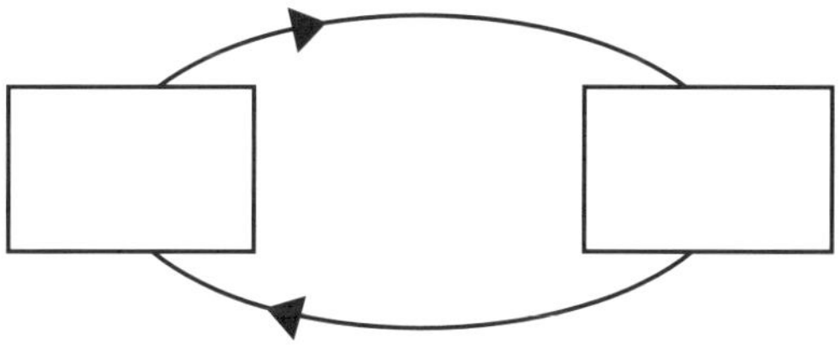

Notes for adults ....

....

Name ______________________________

Which way?

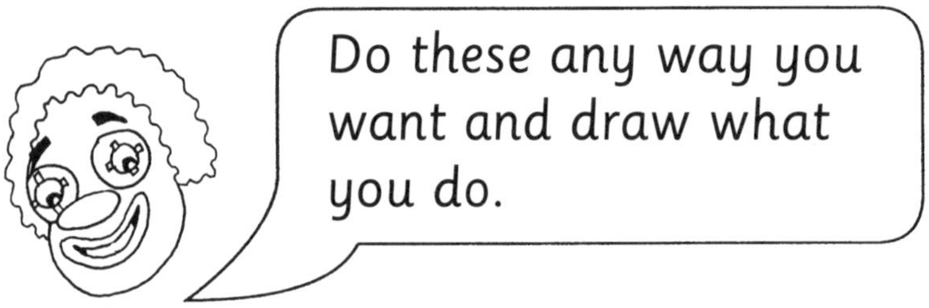

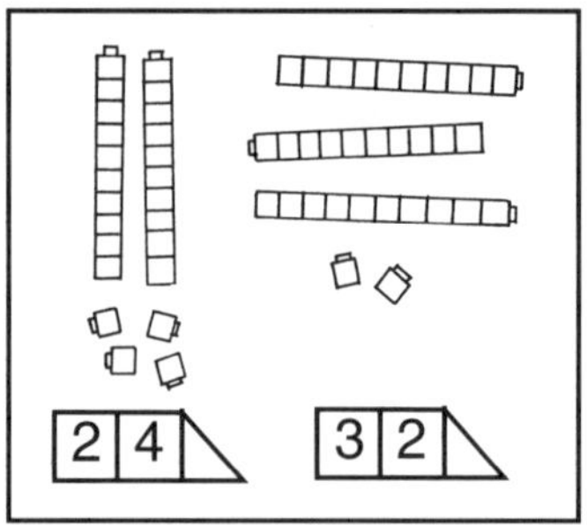

1. I did like this:

2. I can do it this way as well:

Tick the box. I'm sure I'm right – ☐ yes ☐ no ☐ almost

Notes for adults

Name ____________________

Resource sheet 36

Arrays

1. 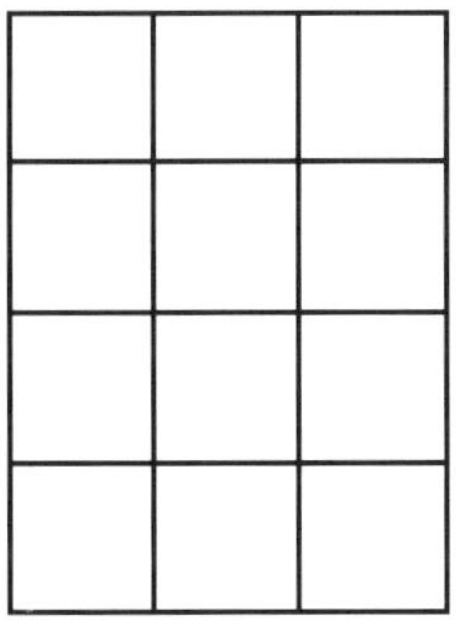

4 rows of ____

is ____ altogether.

4 x ____ = 12

2. 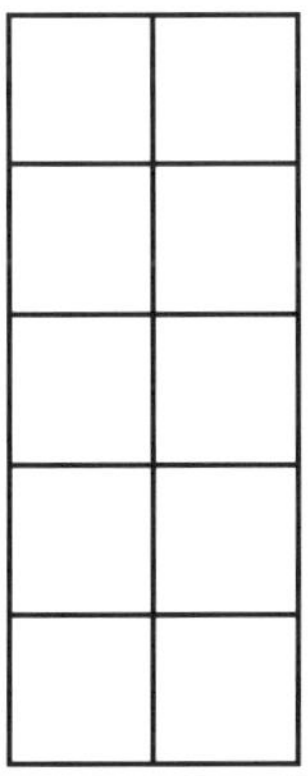

5 rows of ____ is ____

5 x ____ = ____

3. 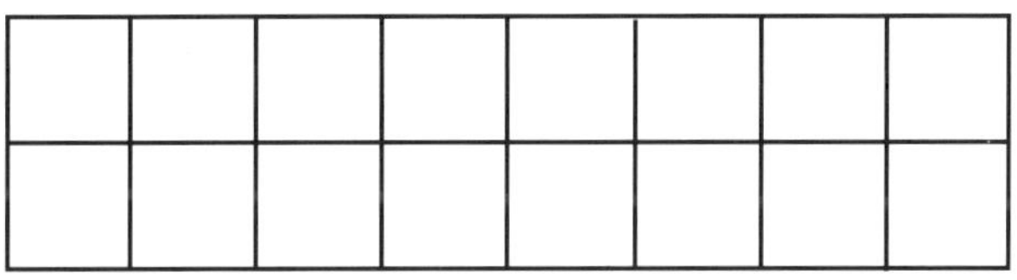

____ rows of 8

is ____ altogether.

2 x 8 = ____

4. 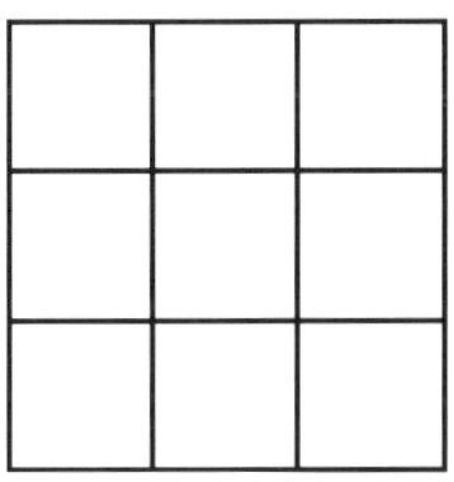

____ rows of ____ is ____ altogether.

____ x ____ = ____

Notes for adults

Make sure the children can use a range of language to describe arrays.

Name ______________________________

Resource sheet 37

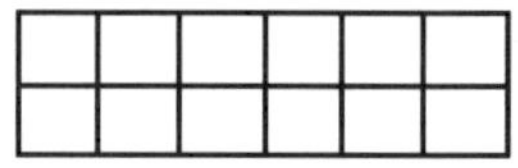

2X6

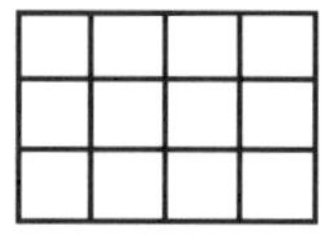

3X4

Drawing arrays

Use ______ cubes.

I made

4 lots of 3

6 groups of 2

times

add

2+2+2+2+2+2

Notes for adults

Name ______________________________

Resource sheet 38

Notes for adults

Name ______________________________

Resource sheet 39

Hopping back – 1

1.

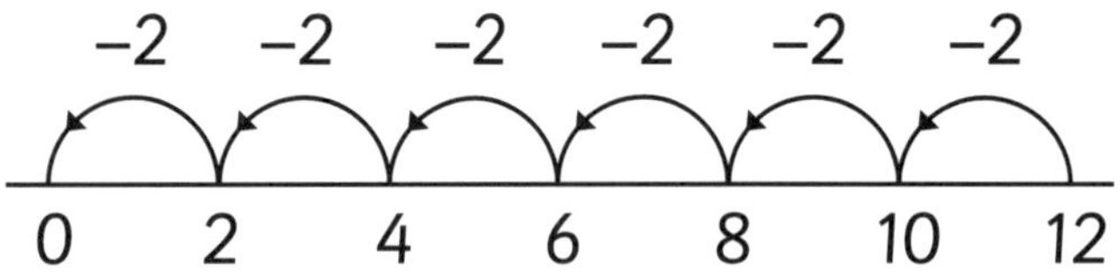

Start on 12. Hop in 2's. This is 6 hops of ☐

12 ÷ 2 = ☐

2.

☐ ☐ ☐ ☐ ☐ ☐

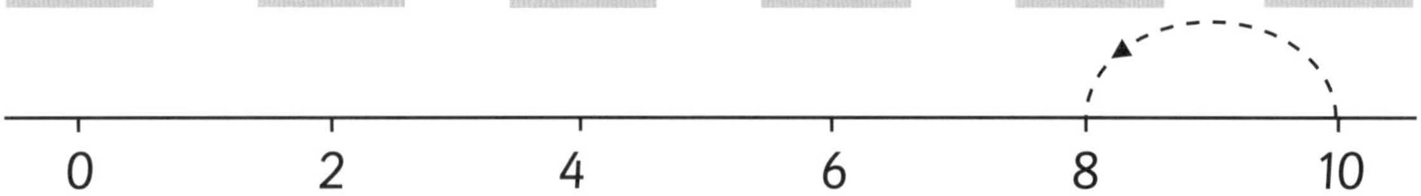

Start on 10. Hop in 2's. This is ☐ hops of ☐

10 ÷ 2 = ☐

3.

☐ ☐ ☐ ☐ ☐ ☐ ☐ ☐

Start on 14. Hop in 2's. This is ☐ hops of ☐

14 ÷ 2 = ☐

Notes for adults

Name ______________________

Resource sheet 40

Hopping back – 2

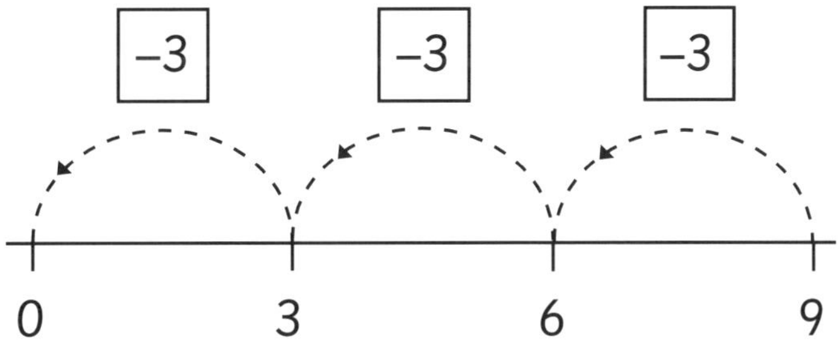

1.

0 1 2 3 4 5 6 7 8 9 10 11 12 13 14 15

Start on ___ Hop in ___s This is ___ hops of ___

___ ÷ ___ = ___

2.

0 1 2 3 4 5 6 7 8 9 10 11 12 13 14 15

Start on ___ Hop in ___s This is ___ hops of ___

___ ÷ ___ = ___

3.

0 1 2 3 4 5 6 7 8 9 10 11 12 13 14 15

Start on ___ Hop in ___s This is ___ hops of ___

___ ÷ ___ = ___

Notes for adults

Name ______________________________

Share it out – 1

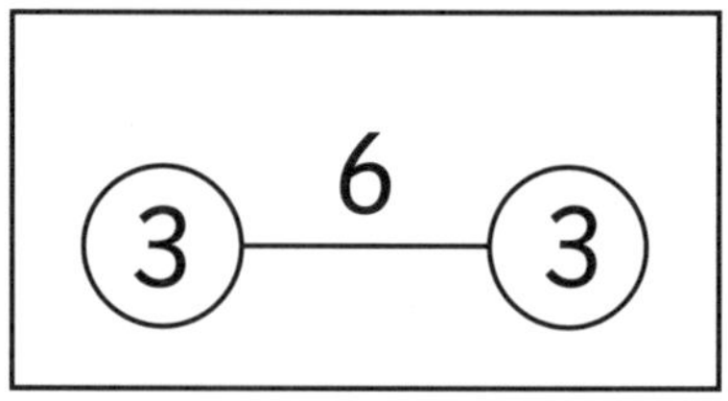

1.

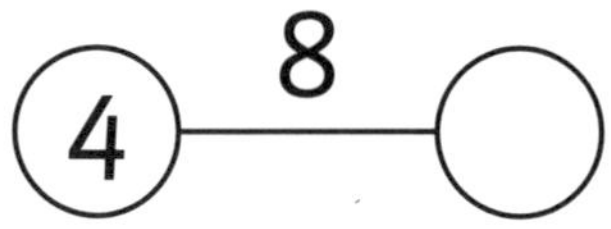

2.

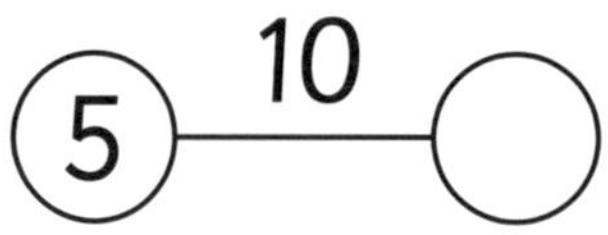

3.

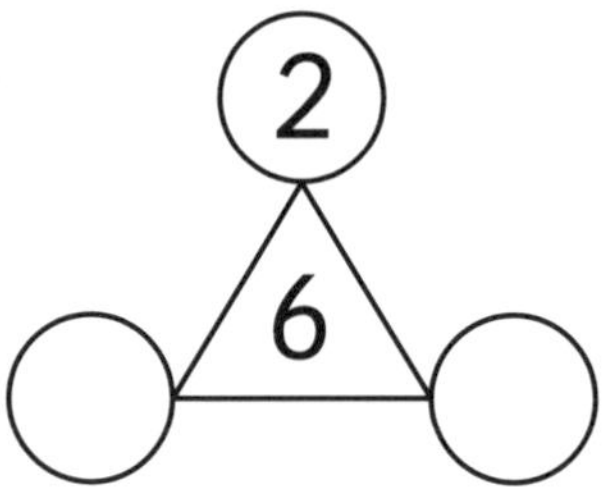

4.

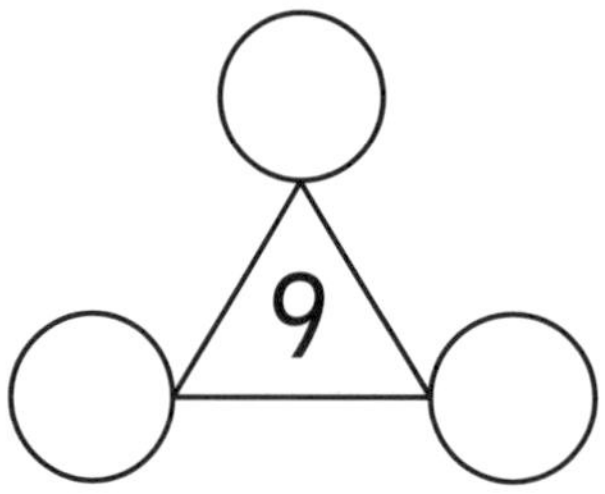

5.

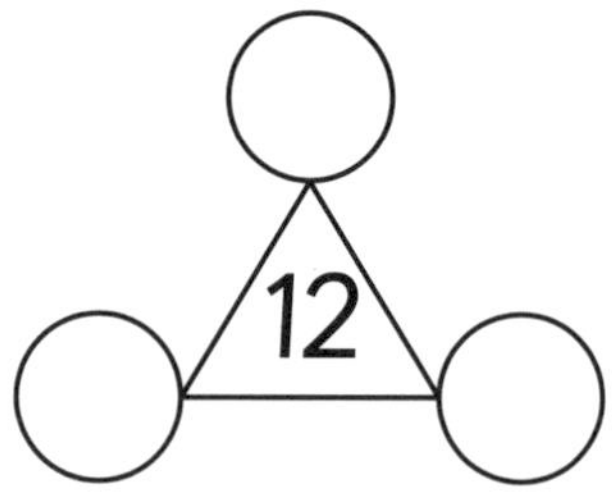

6.

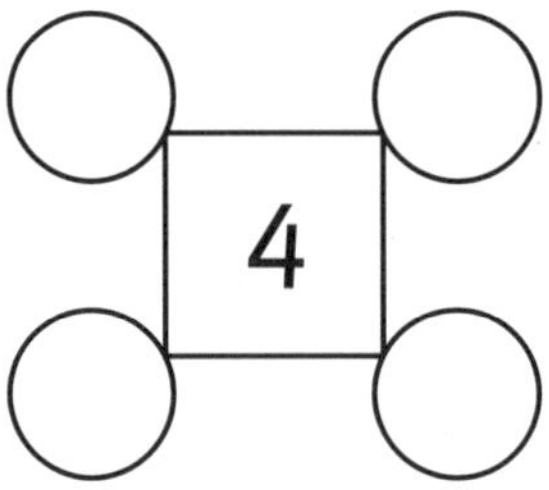

7.

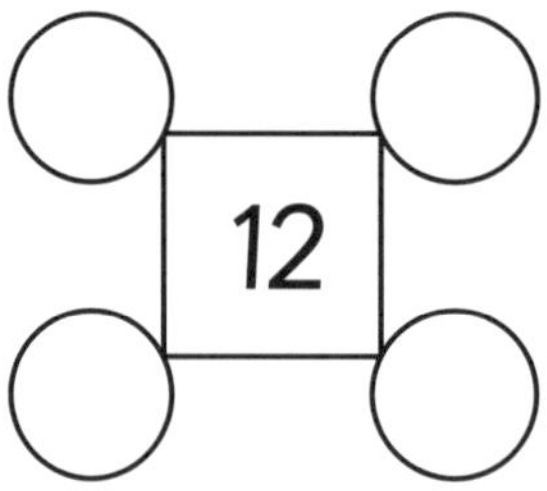

8.

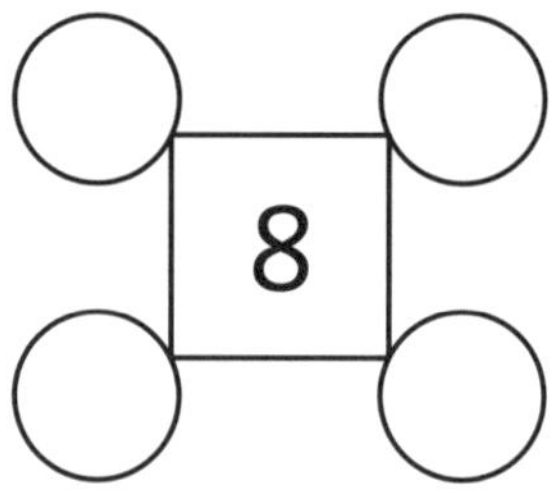

9.

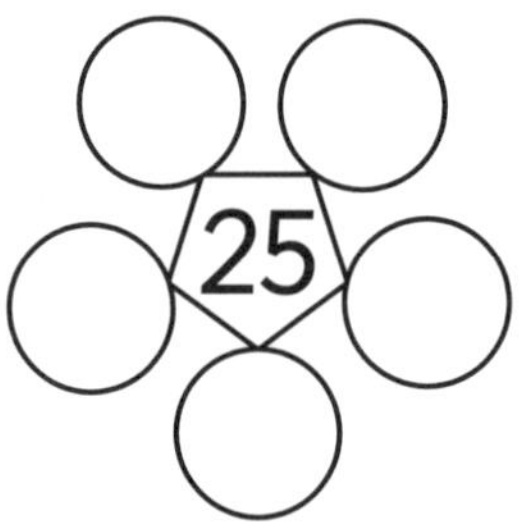

10.

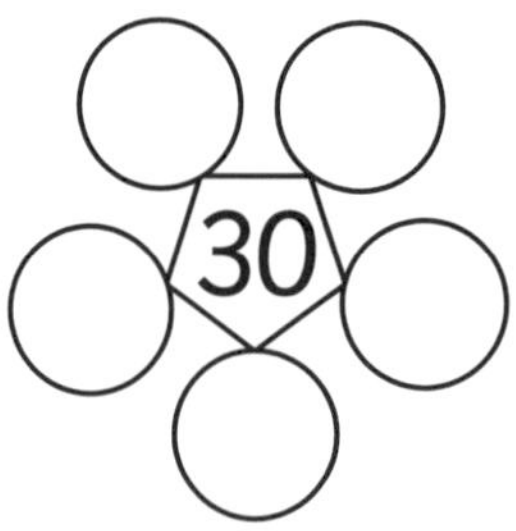

Notes for adults

Share out the numbers in the middles equally.
There is space at the bottom for the child to make their own shape.

Name ______________________________

Share it out – 2

4 apples. 2 each.

1. 

2. 

3. 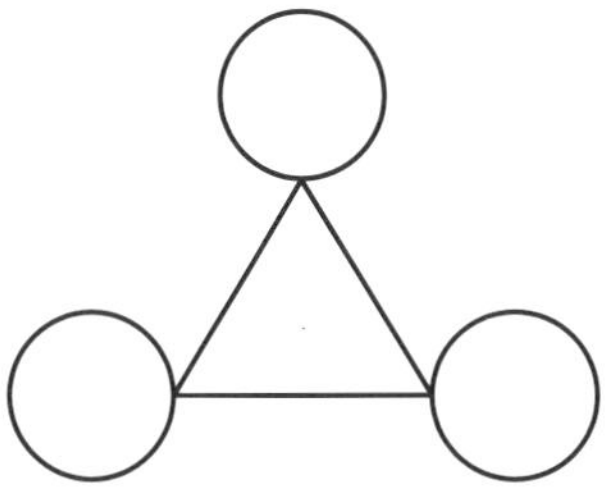

4. 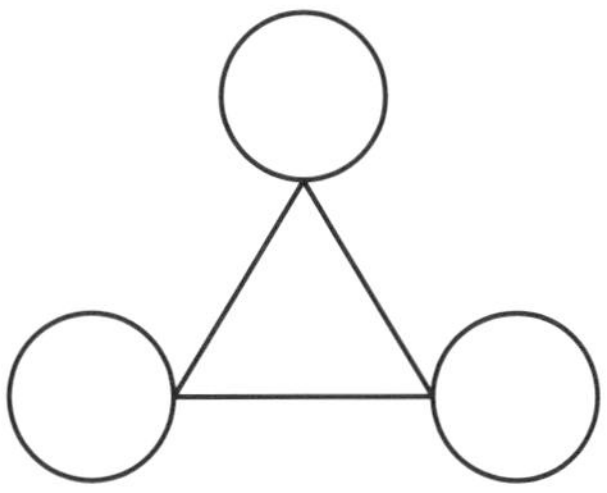

5. 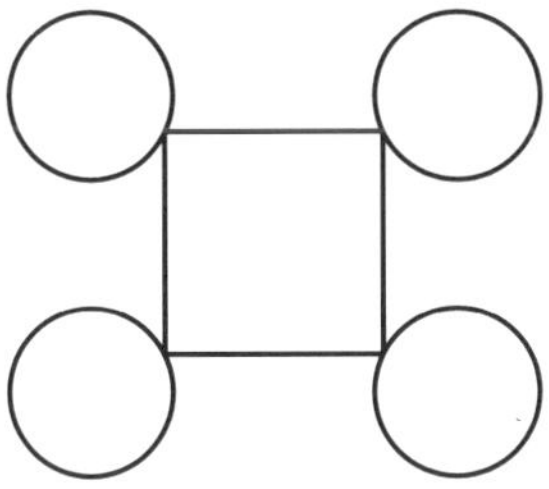

6. 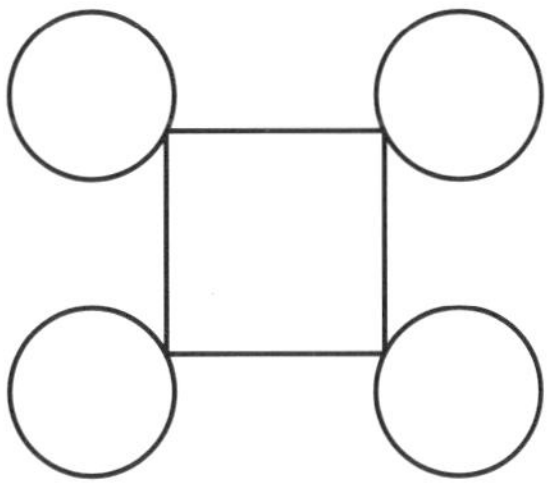

7. 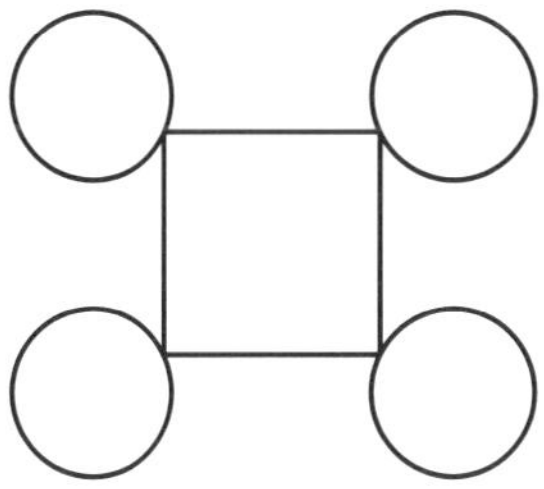

8. 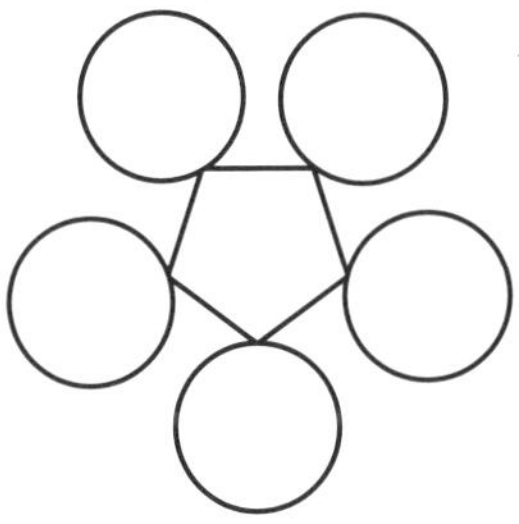

9. 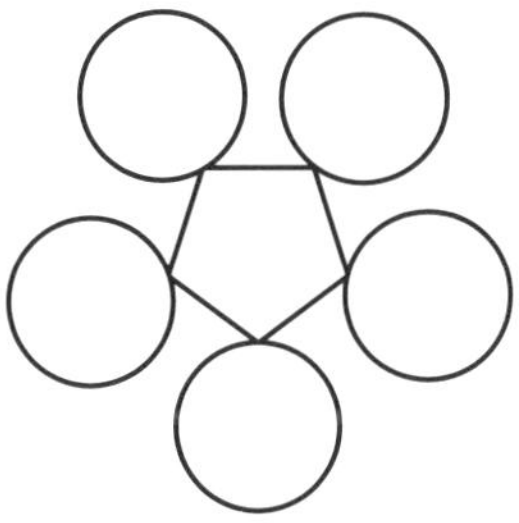

Notes for adults

Name ______________________________

Resource sheet 43

Grab for gold

13 → 6, 6 r 1

1. ☐ cubes
○ ○ r ☐

2. ☐ cubes
○ ○ r ☐

3. ☐ cubes
○ ○ r ☐

4. ☐ cubes
○ ○ r ☐

5. ☐ cubes
○ ○ ○ r ☐

6. ☐ cubes
○ ○ ○ r ☐

7. ☐ cubes
○ ○ ○ r ☐

8. ☐ cubes
○ ○ ○ r ☐

Notes for adults ....

Name ______________________________

Share the biscuits

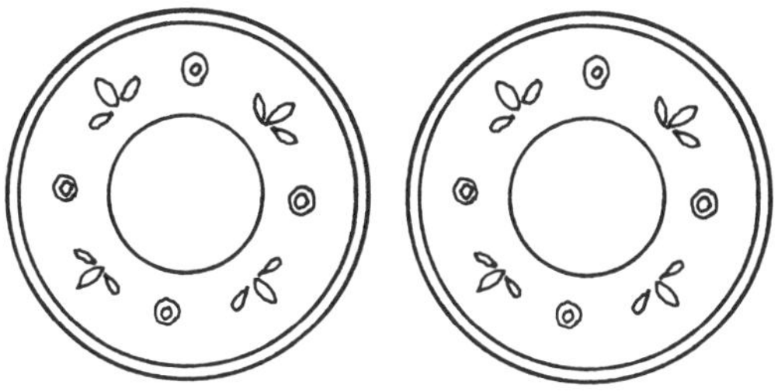

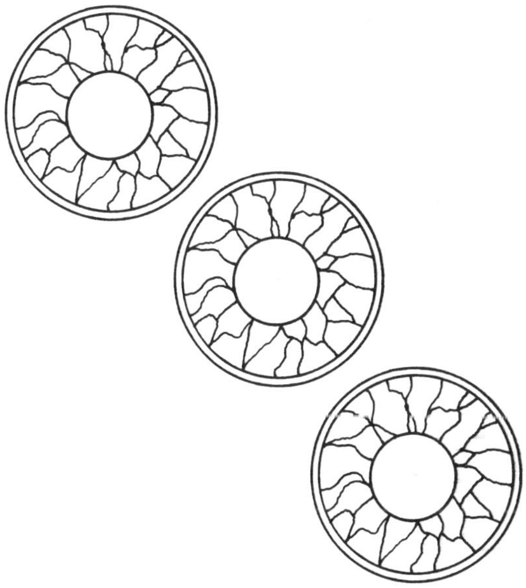

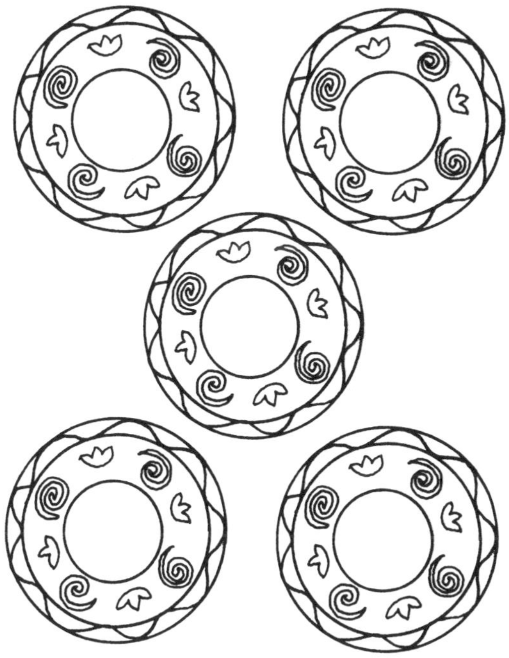

Notes for adults

Fraction cards

$\frac{1}{2}$	$\frac{1}{4}$	$\frac{3}{4}$	$\frac{1}{5}$
$\frac{1}{10}$	$\frac{2}{10}$	1 whole	$\frac{2}{5}$
$\frac{5}{10}$	$\frac{2}{4}$	$\frac{3}{10}$	$\frac{4}{10}$
$1\frac{1}{2}$	$1\frac{1}{4}$	$1\frac{1}{10}$	$1\frac{1}{5}$

Pizzas and cakes – 1

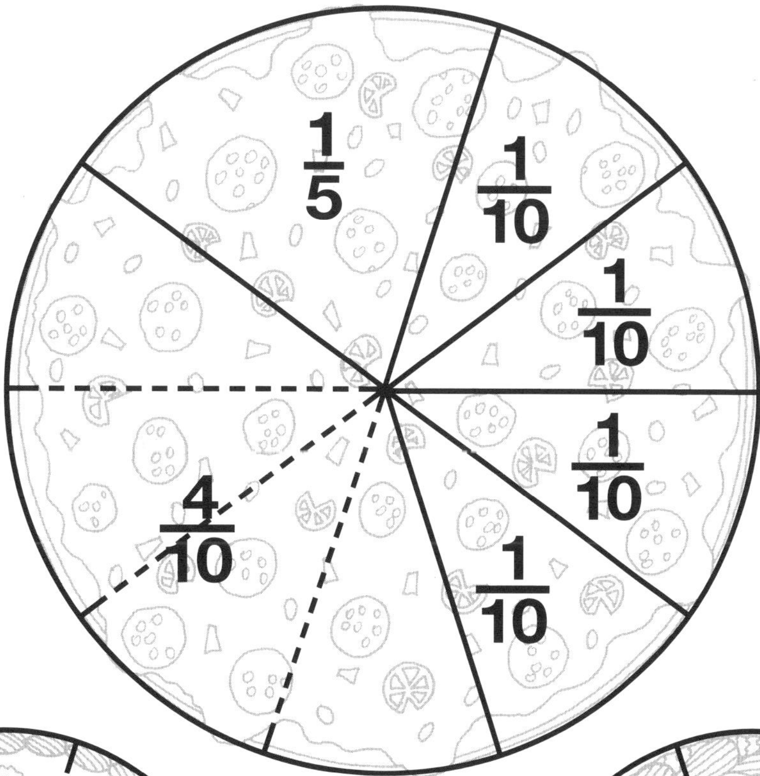

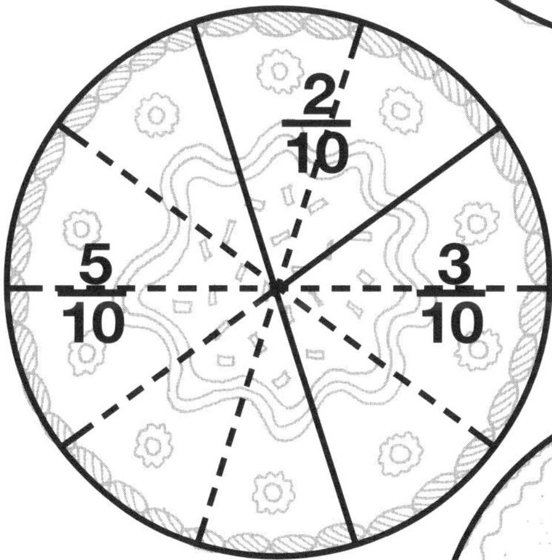

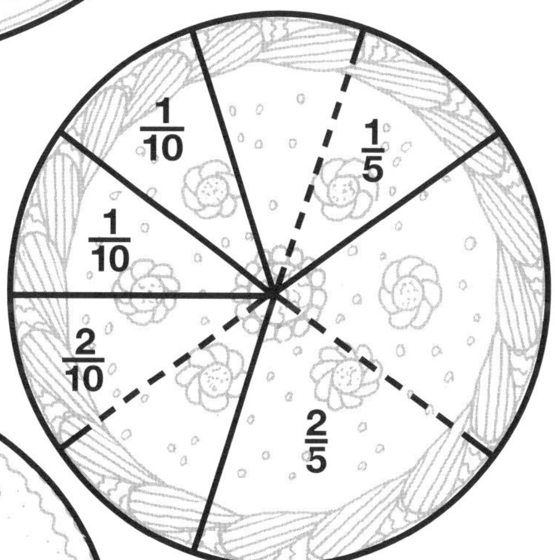

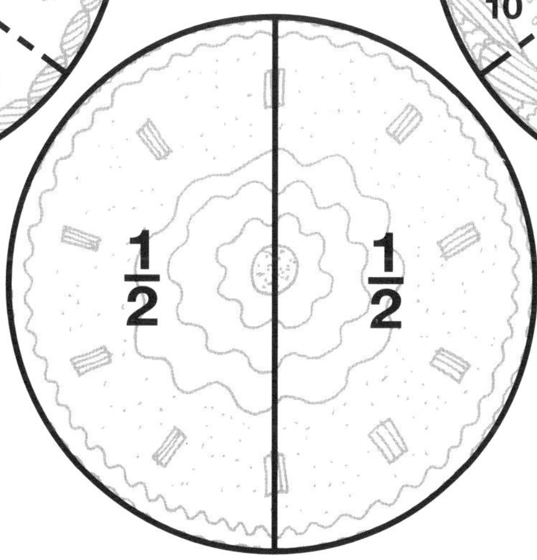

Notes for adults

Pizzas and cakes – 2

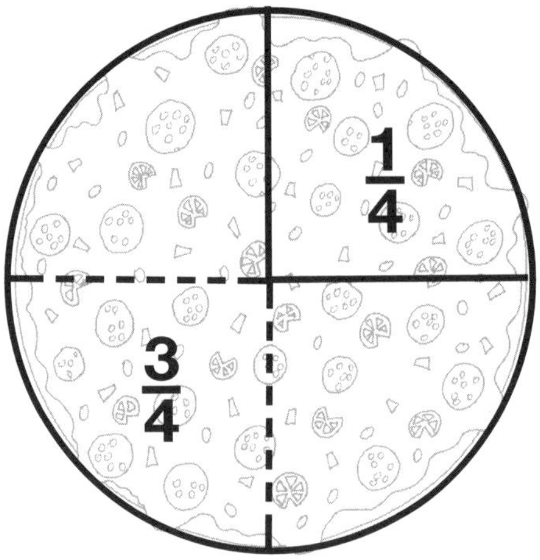

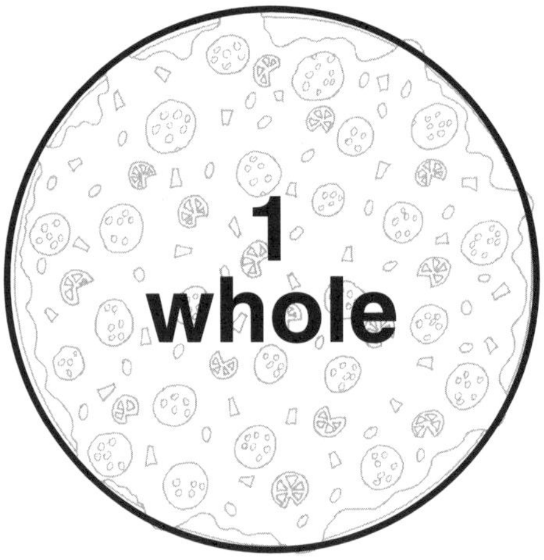

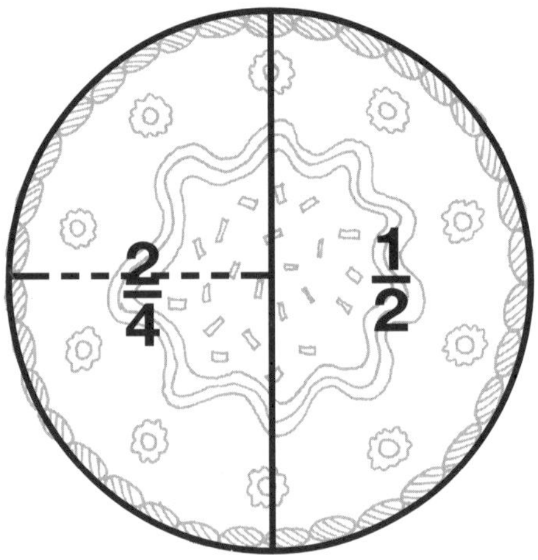

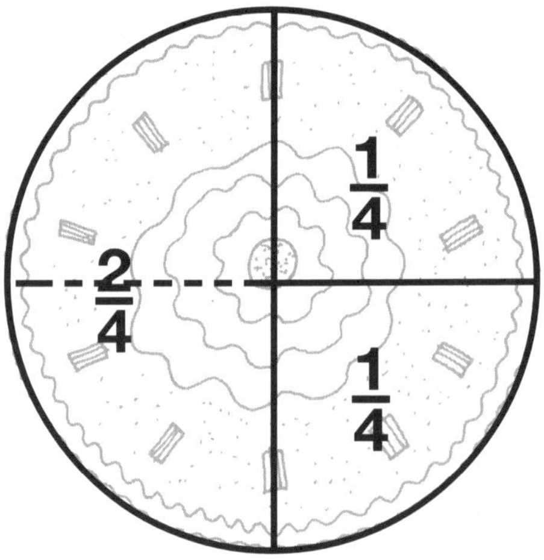

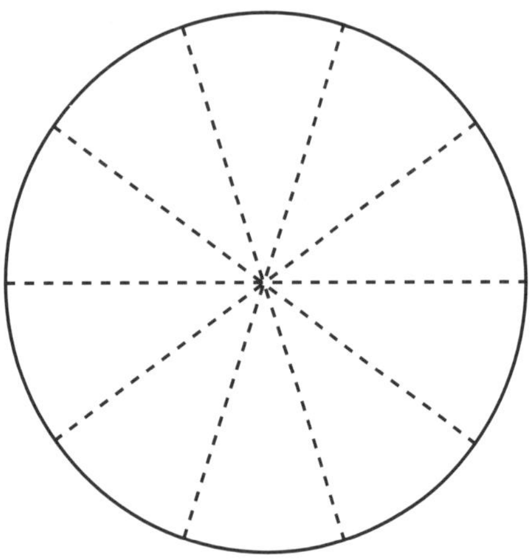

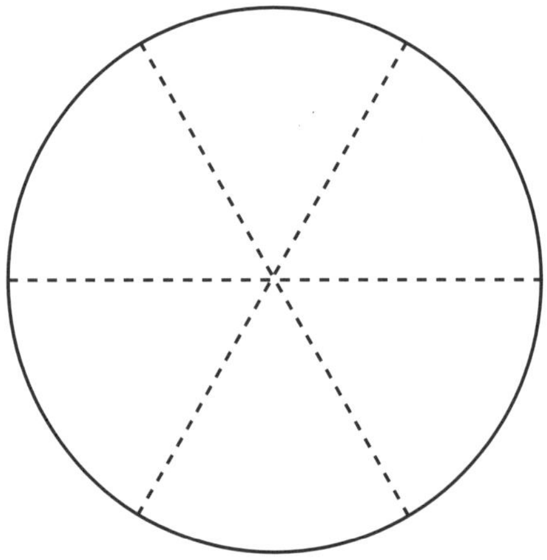

Notes for adults

Name ______________________________

Resource sheet 48

Fractions of numbers

1. Circle $\frac{1}{2}$.

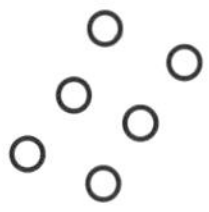

$\frac{1}{2}$ of 6 is

2. Circle $\frac{1}{2}$.

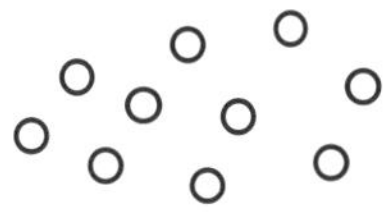

$\frac{1}{2}$ of 10 is

3. Find $\frac{1}{4}$.

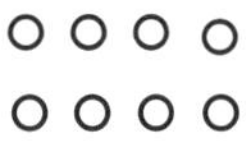

$\frac{1}{4}$ of 8 is

4. Find $\frac{1}{4}$.

$\frac{1}{4}$ of 12 is

Use cubes.

5. $\frac{1}{2}$ of 16 is

6. $\frac{1}{2}$ of 20 is

7. $\frac{1}{2}$ of 18 is

8. $\frac{1}{2}$ of 22 is

9. $\frac{1}{4}$ of 16 is

10. $\frac{1}{4}$ of 20 is

Use 12 24 48 or 96

5. Find $\frac{1}{2}$ and $\frac{1}{4}$.

Notes for adults

Bingo cards

Notes for adults

Name ____________________

Resource sheet 50

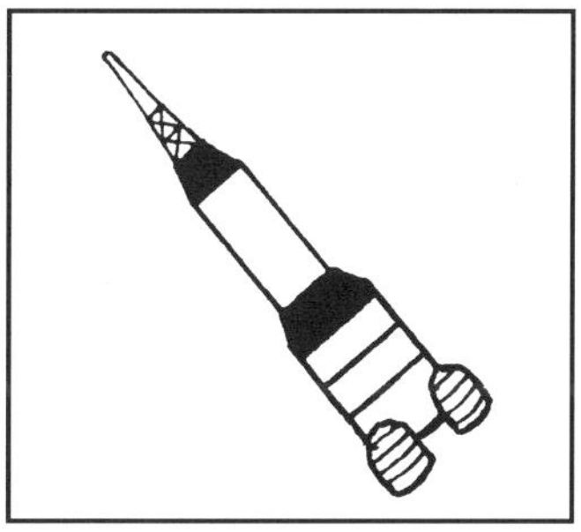

Tenths and fifths

1. Circle $\frac{1}{10}$

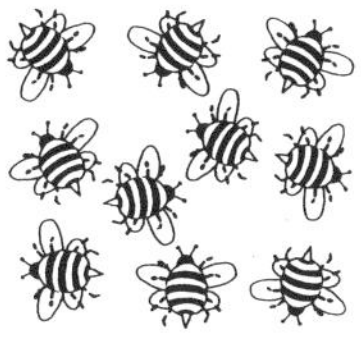

2. Circle $\frac{1}{10}$

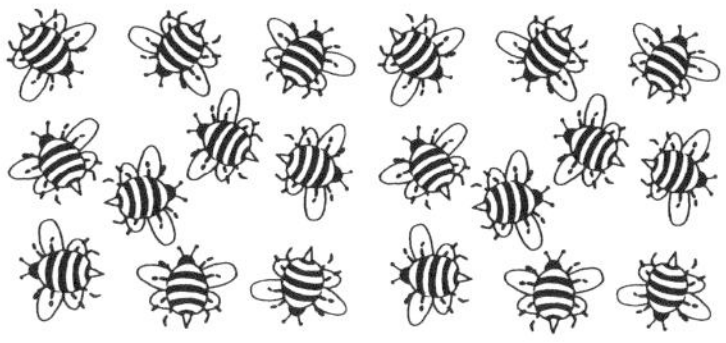

$\frac{1}{10}$ of 10 is ____

$\frac{1}{10}$ of 20 is ____

3. $\frac{1}{10}$ of 50 is ____

4. $\frac{1}{10}$ of 100 is ____

5. $\frac{1}{10}$ of 60 is ____ so $\frac{1}{5}$ of 60 is ____

6. $\frac{1}{10}$ of ____ is ____ so $\frac{1}{5}$ of ____ is ____

7. $\frac{1}{10}$ of ____ is ____ so $\frac{1}{5}$ of ____ is ____

Choose other numbers.

Notes for adults

Name ______________________________

Decimals and fractions

1. Key in 3 ÷ 10 = ☐

2. Key in 4 ÷ 10 = ☐

Put the numbers on the number line.

3.

0	$\frac{1}{10}$	$\frac{2}{10}$	$\frac{3}{10}$	$\frac{4}{10}$	$\frac{5}{10}$	$\frac{6}{10}$	$\frac{7}{10}$	$\frac{8}{10}$	$\frac{9}{10}$	1
zero	0.1	0.2			0.5					1

4.

Make 0.5

5 ÷ 10

6 ÷ 12

Notes for adults

Name ______________________________

Fractions of shapes

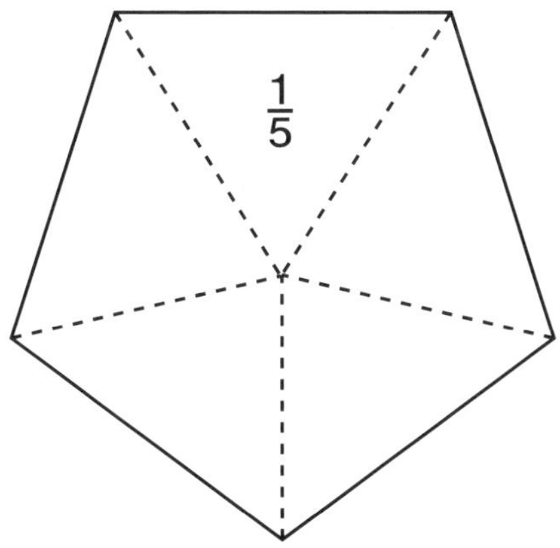

regular pentagon

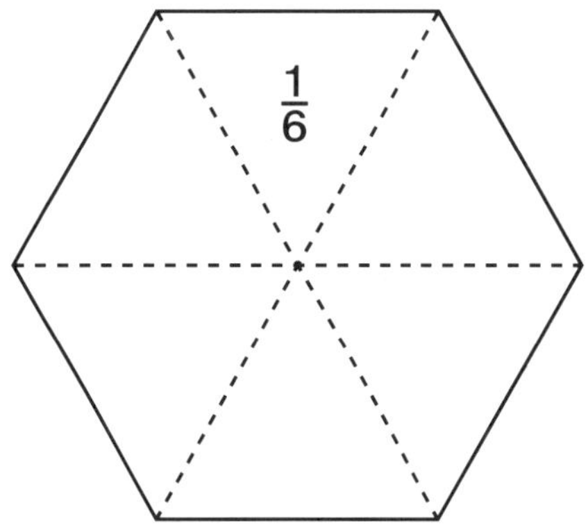

regular hexagon

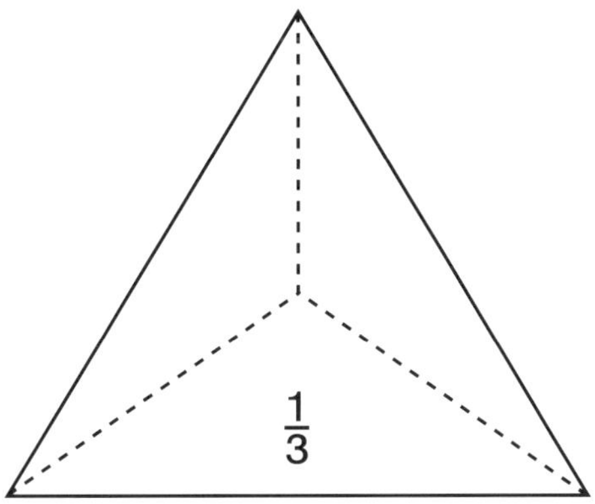

equilateral triangle

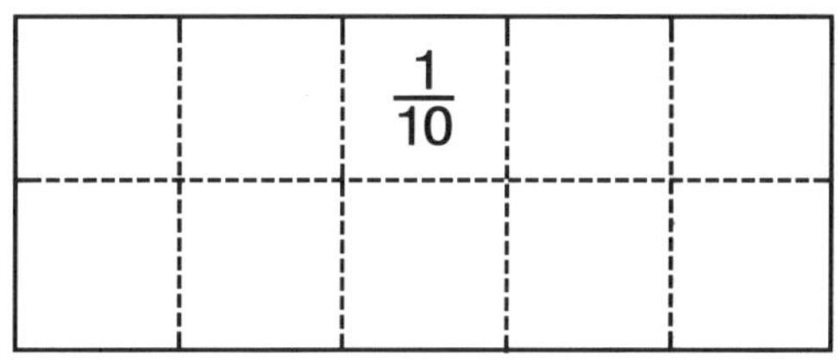

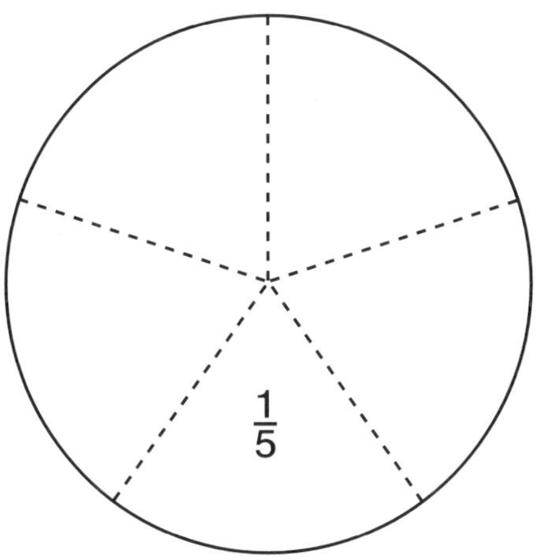

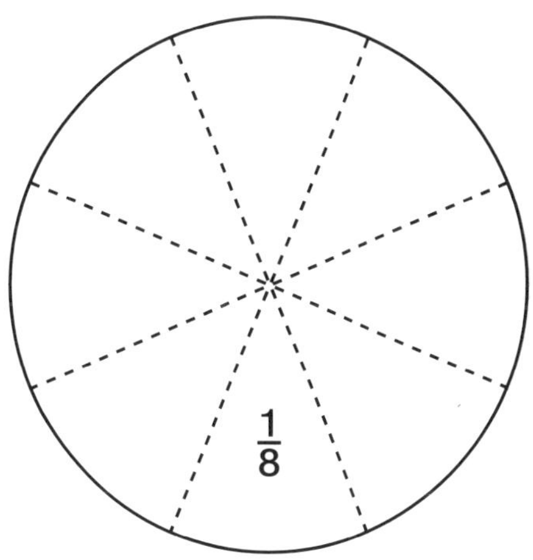

Notes for adults

Name ______________________________

At the shop

1. Spend [ ] in the shop.

Write what you buy.

Costs [ ]

Change [ ]

2. Mum gives you [ ] to buy presents for your party.

Write what you buy.

Costs [ ]

Change [ ]

Notes for adults

Name ______________________

Stories

1. 10 ugly bugs came to the ball.
   3 went home early.
   How many stayed late?

   10 – ☐ = ☐

2. Each bug bus arrived with 10 bugs in each one. There were 8 buses. How many bugs?

3. By midnight, there were 16 ladybirds, 18 grasshoppers and 10 flies all dancing. How many dancers?

4. There are 16 packets of crisps. How many bugs can have 2 each?

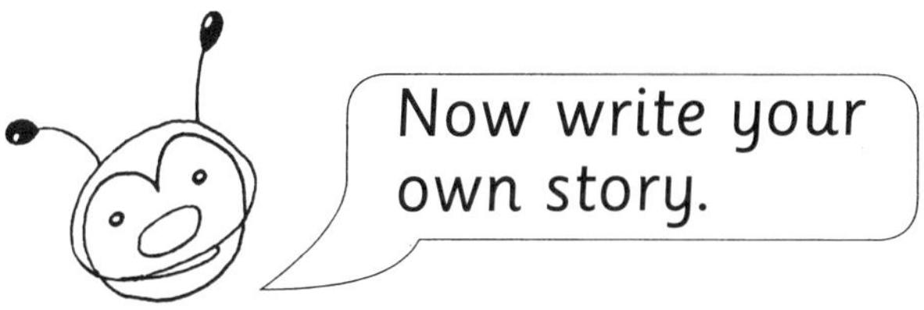

Notes for adults

You might need to read these problems to the children.

Money game – 1

Resource sheet 55

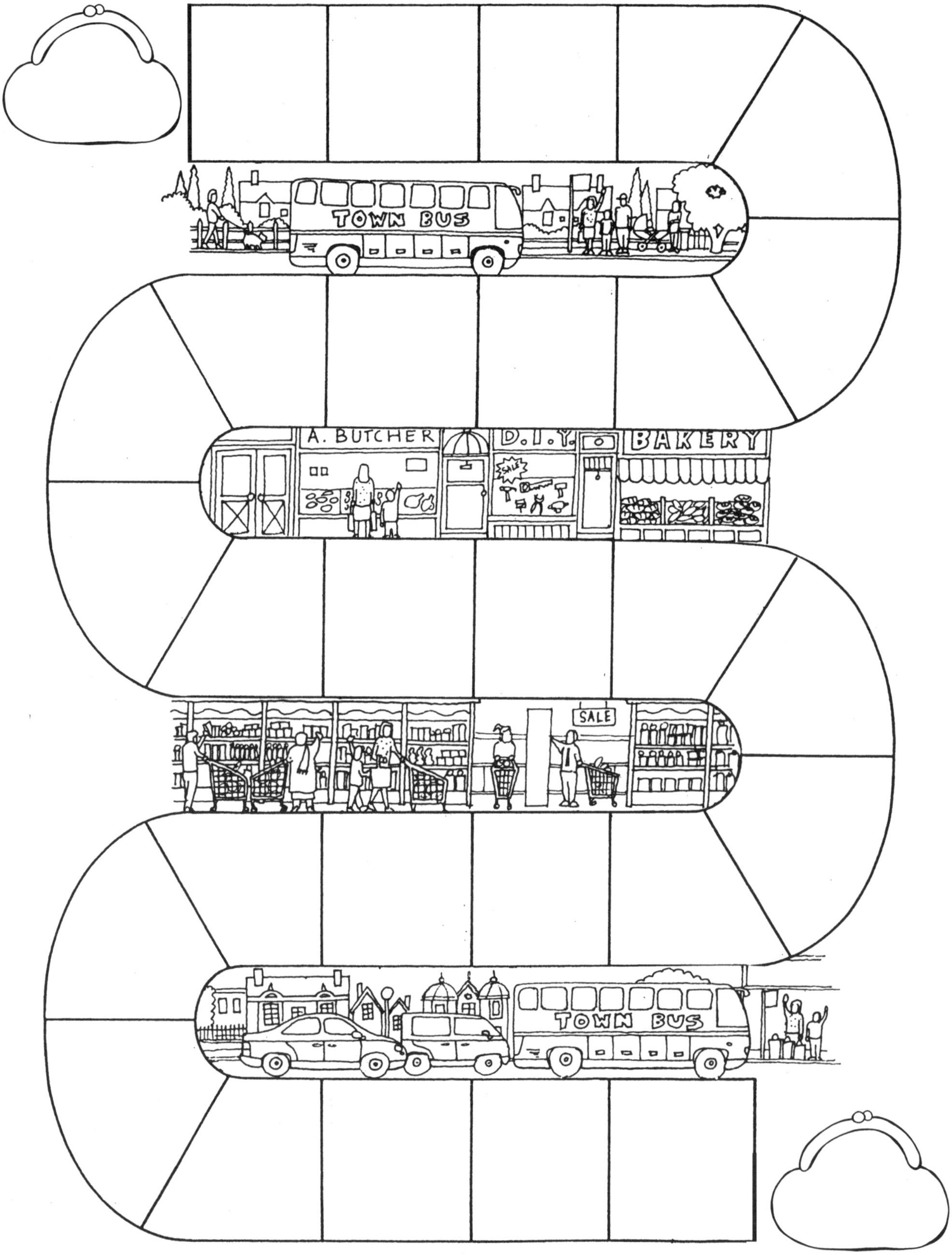

Notes for adults

Money game – 2

Resource sheet 56

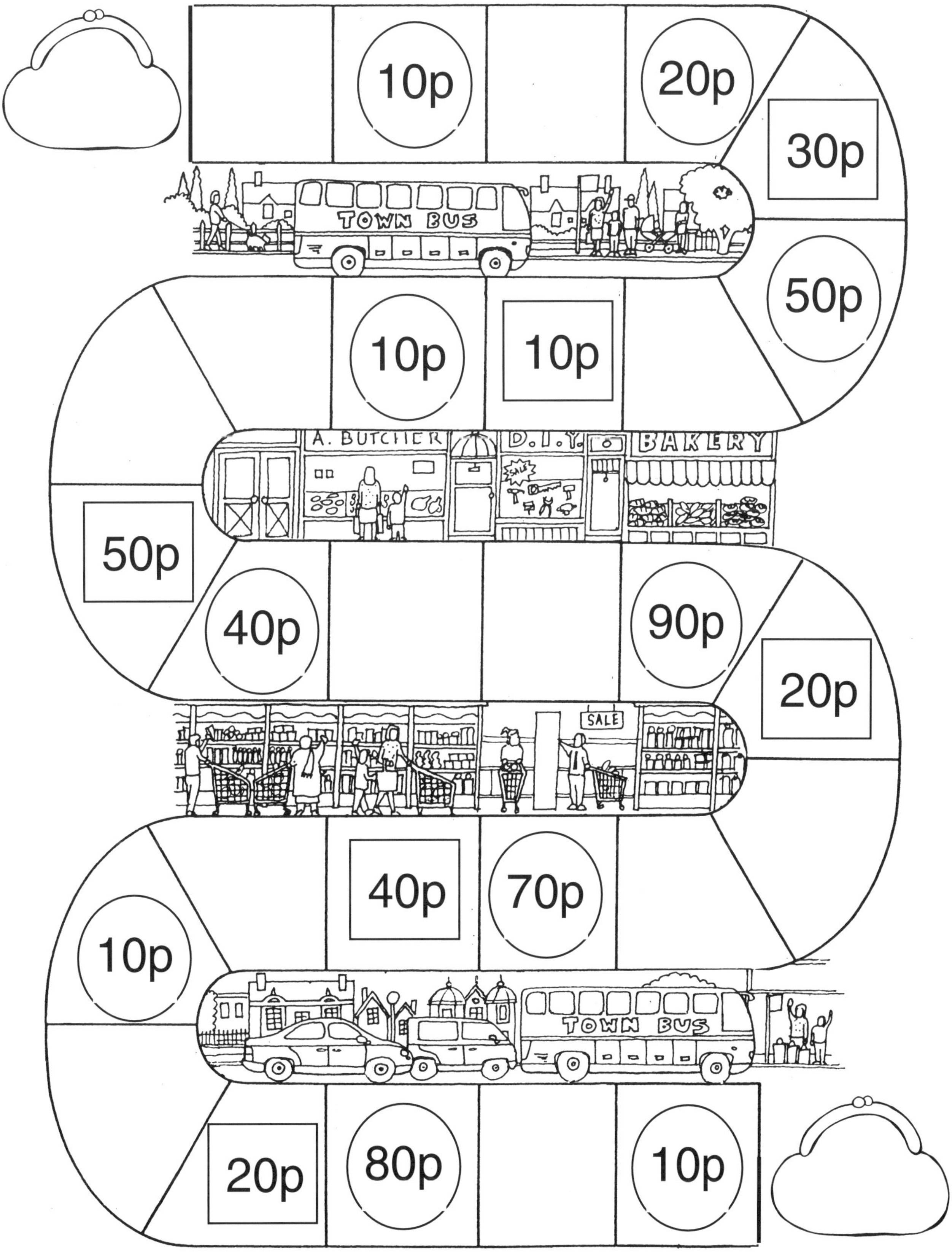

Notes for adults

You might need to read these amounts to the children.

Name ______________________________

Money problems

1. 

10 cost __________

2. 

10 cost __________

3. 

10 cost __________

4. Bus fares

	number of people		
	1	10	100
zoo 15p each	15p		
seaside 50p each	50p		

5. 

How much for 1 box? __________

How much for 5 boxes? __________

How much for 20 boxes? __________

Notes for adults

Name ______________________________

Buses and bags

1. 17 on the bus. 6 get off. Put a ✔ by the correct number sentence.

17 + 6 ☐   17 ÷ 6 ☐   17 – 6 ☐

2. 13 cakes in a bag. Eat 5. Is this

13 – 5 ☐ , 13 + 5 ☐ or 13 x 5 ☐ ?

3. 21 marbles in a bag. Lose 8. Is this

21 x 8 ☐ , 21 + 8 ☐ or 21 – 8 ☐ ?

4. 12 get on the bus and 3 more get on.

12 x 3 ☐ , 12 + 3 ☐ or 12 – 3 ☐ ?

5. The baker has 6 bags of 5 cakes. How many cakes?
Is this 6 – 5 ☐   6 + 5 ☐   6 x 5 ☐ ?

6. 200 children go to the museum. Each bus takes 50 children. How many buses are needed? Is this

200 – 50 ☐ , 200 ÷ 50 ☐ or 200 x 50 ☐ ?

Notes for adults

Provide the children with operation cards.

Name ______________________________

Resource sheet 59

Lots of legs

Draw sets of 24 legs.
Write number sentences.

1. ________ = 24

2. ________ = 24

3. ________ = 24

4. ________ = 24

Notes for adults ....

Name ______________________________

What's my number?

1. 

I have a number in my head.
____ + 4 = 10

6 + 2 = 4 + 4
24 = 8 + 8 + 8
12 – 2 = 5 + 5

2. ____ + 5 = 12

3. ____ – 6 = 10

4. 12 + ____ = 18

5. 14 + ____ = 19 + 1

6. ____ x 2 = 20

7. 18 ÷ ____ = 9

8. Half of ____ is 12.

9. half of ____ is 25.

10. Double ____ is 100.

11. Double ____ is 60.

Now make up some for a friend.

Notes for adults

You might find it helpful to provide a balance.

Name ______________________________

Bugs at the ball

1.  drinks 20p

10 cost

5 cost

2.  toffee apples 30p

10 cost

5 cost

3.  cakes 15p

10 cost

5 cost

4.  lollies 50p

10 cost

5 cost

5. You choose things.

cost

10 cost

5 cost

Notes for adults

Remind the children how to multiply by 10, then halve that amount.

Name ______________________________

Bugs on buses

1

2

3

Now you make up numbers.

4

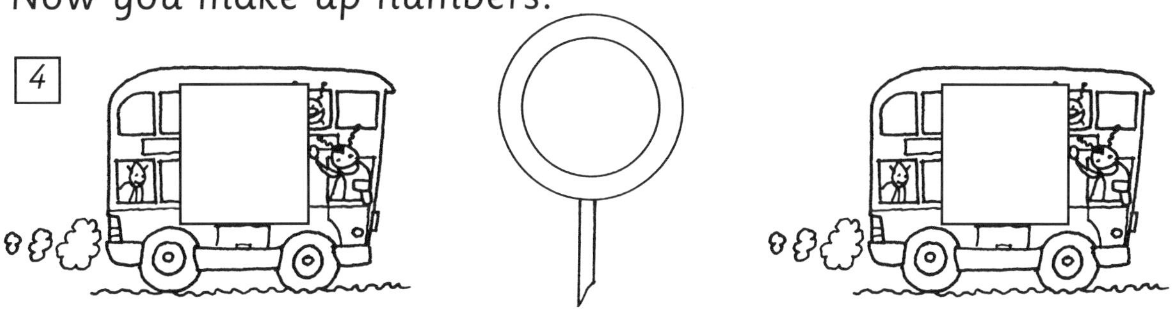

Notes for adults

Name ______________________________ Resource sheet 63

More 'What's my number?'

1. 

I have a number in my head.

I double it and add 1.

(☁ x 2) + 1 = 13

2. (☁ + 3) x 2 = 12

3. (☁ – 4) x 2 = 8

4. (☁ – 1) + 3 = 10 + 4

5. 12 + 6 = ☁ – 1

6. 27 – 9 = ☁ + 3

7. 36 + 2 = ☁ – 5

8. *Find different ways to solve this.*

☁ + □ = ○ + 1

Notes for adults

Remind the children to do the brackets first.

Name ______________________________

Resource sheet 64

Tens and hundreds

Get brilliant at multiplying by 10!

	Guess	Check	✔
13 x 10	130	130	✔
26 x 10			
19 x 10			
39 x 10			

Now divide by 10.

	Guess	Check	✔
390 ÷ 10			
260 ÷ 10			
570 ÷ 10			
620 ÷ 10			

Now choose numbers and x by 100.

Notes for adults

Make sure the children are seeing the pattern.

Name ______________________

What goes in the box?

12 [x] 3 = 36   49 [÷] 7 = 7

1. 48 □ 2 = 24

2. 360 □ 2 = 180

3. 127 □ 26 = 101

4. 118 □ 18 = 136

5. 29 + 31 = 15 □ 45

6. 51 + 49 = 25 □ 75

7.

8.

9.

10.

Notes for adults

Encouraging children to investigate what a calculator can do can help them to understand maths more easily.

Name ______________________________

Measuring units

1. 1 kilometre = 1000 metres

$\frac{1}{2}$ kilometre = __________ metres

$\frac{1}{10}$ kilometre = __________ metres

10 kilometres = __________ metres

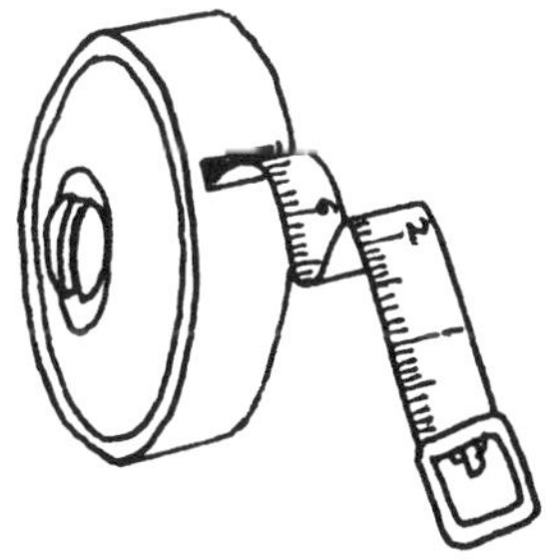

2. 1 metre = __________ centimetres

$\frac{1}{2}$ metre = __________ centimetres

$\frac{1}{10}$ metre = __________ centimetres

10 metres = __________ centimetres

1cm = __________ millimetres

3. Write another table.

1 kilogram = ______________________

$\frac{1}{2}$ kilogram = ______________________

$\frac{1}{10}$ kilogram = ______________________

10 kilograms = ______________________

Notes for adults

Let the children use calculators to explore the larger numbers.

Name ______________________________

Measuring problems

1. How tall is Ishal?

2. What length is shown?

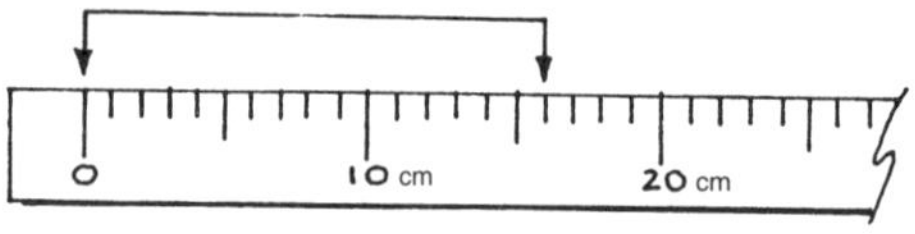

3. How heavy is the parcel?

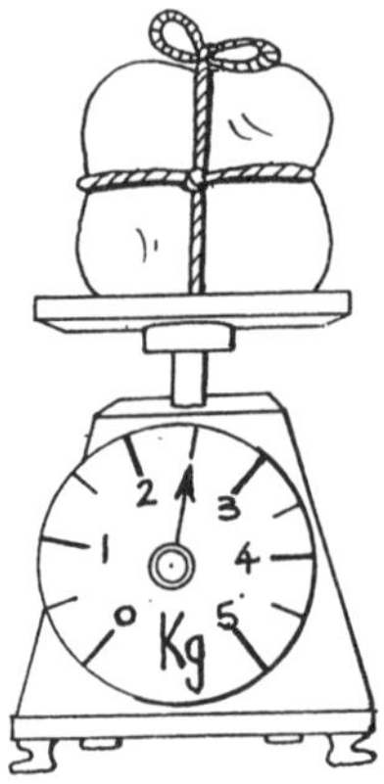

4. How much is in the jugs?

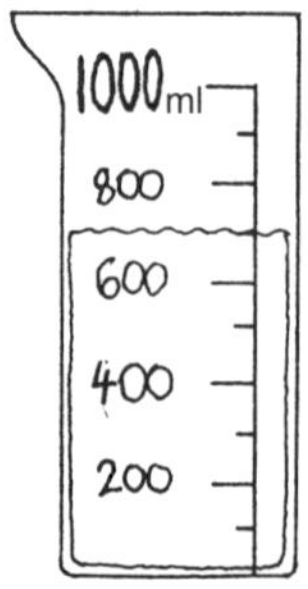

Notes for adults

Name ______________________________

Round up or down?

1.

Tickets cost £5

Jake has £36

☐ Round up.

☐ Round down.

He can buy ________ tickets.

2.

Sally wants 100 metres of bunting.

☐ Round up.

☐ Round down.

She needs __________ packs.

3.

Each minibus takes 12 people.

☐ Round up.

☐ Round down.

For 51 people you need ____________ minibuses.

4.

20 hats in a box.
84 people need a hat.

☐ Round up.

☐ Round down.

How many boxes are needed? ____________ boxes

Make up a rounding down story with money.

Notes for adults

Clip art

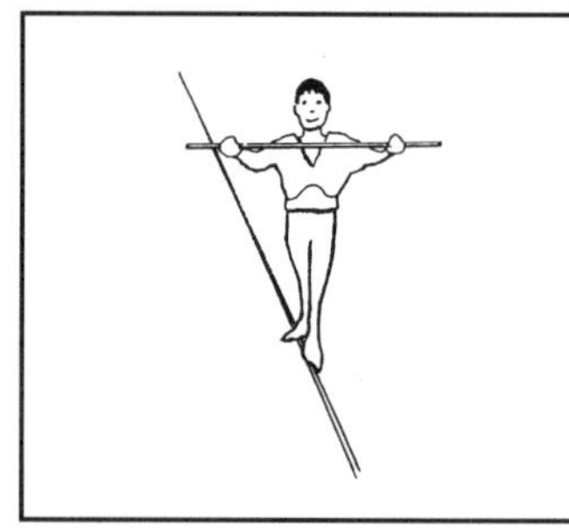

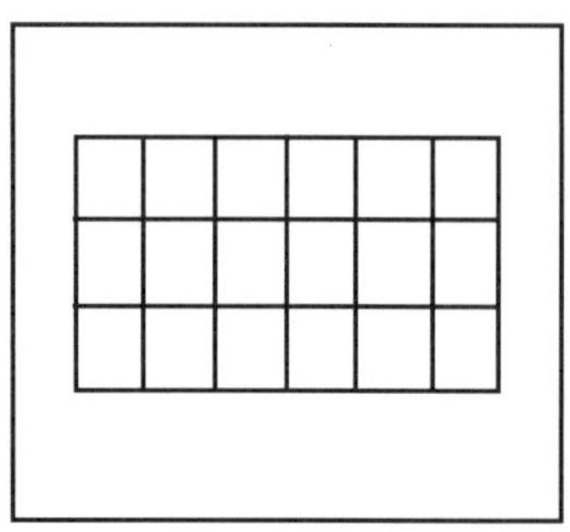
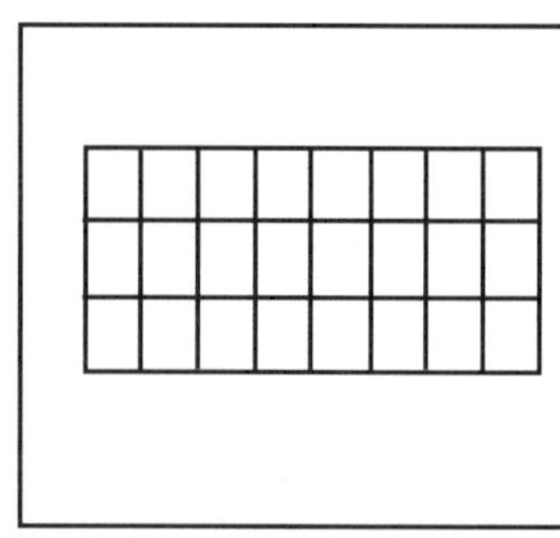

Clip art

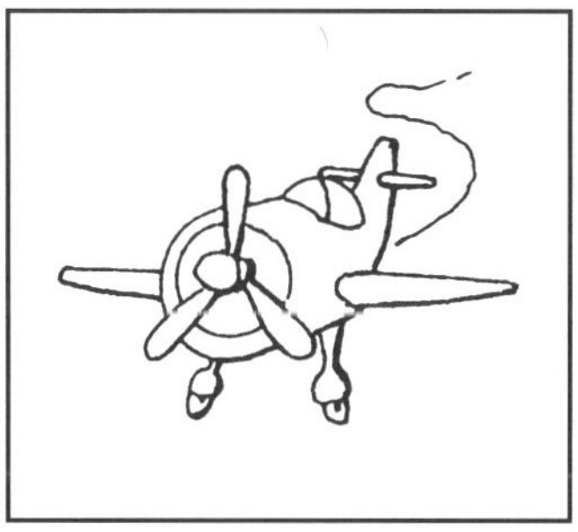
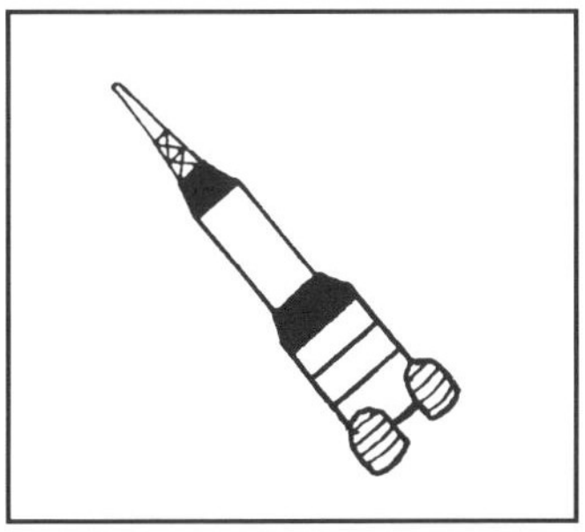

Clip art

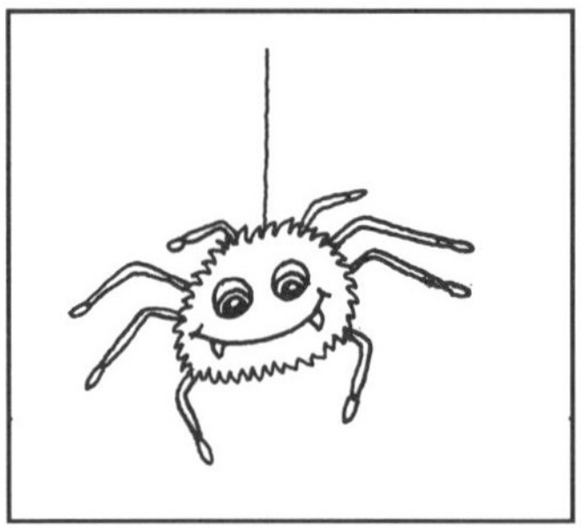
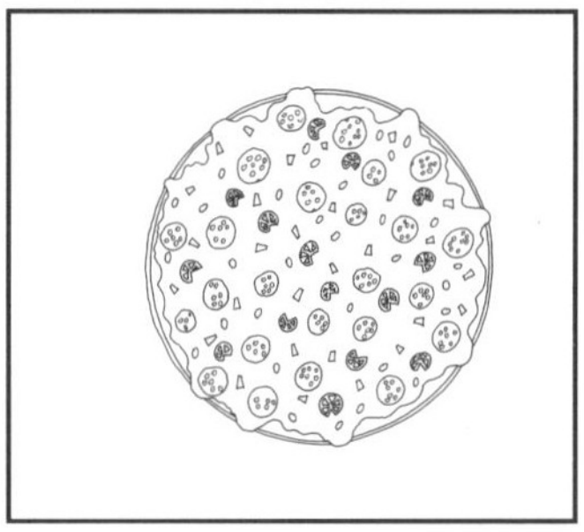

List of Assessment Focuses

Assessment focus	Chapter	Date achieved / comments
Can the child order numbers to 100, recognise the pattern of the 10's and 1's and read number names?	1	
Can the child order numbers? Do they know the value of each digit? Can they use the language of 'more than' and 'less than'?	1	
Can the child understand 'more than' and 'less than'?	1	
Can the child count on and back in 10's and 100's starting from any two- or three-digit number?	1	
Can the child give sensible estimates of numbers of objects to 100, check count by grouping in 10's, and round two digit numbers to the nearest 10 (then to the nearest 100)?	1	
Can the child recognise and extend number sequences, stating the rule?	2	
Can the child recognise odd and even numbers?	2	
Can the child recognise multiples of 2, 5 and 10 (then 3 and 4)?	2	
Can the child add two, three and four small numbers together mentally?	3	
Can the child use a range of language for addition and subtraction?	3	
Can the child add (and subtract) a near multiple of 10 on a number line?	3	
Can the child put the larger number first when adding numbers?	3	
Can the child find pairs of numbers to 100?	3	
Can the child work out a missing number in addition and subtraction sentences?	4	
Can the child remember number bonds to 10 and up to 20?	4	
Can the child use corresponding addition and subtraction facts?	4	
Can the child understand that if you count back, then forward the same number, you end up where you started?	4	
Can the child add two two-digit numbers?	4	

Assessment focus	Chapter	Date achieved / comments
Can the child describe arrays using the language of multiplication?	5	
Can the child do repeated subtraction on a number line?	5	
Can the child divide into equal groups (and work out remainders)?	5	
Can the child derive quickly division facts that relate to multiplication tables?	5	
Can the child recognise simple fractions that are several parts of the whole?	6	
Can the children recognise simple equivalent fractions and order them on a number line?	6	
Can the child find fractions of numbers, extending to: can the child find tenths of numbers and work out how to find fifths?	6	
Can the child link decimals and fractions?	6	
Can the child recognise fractions that are several parts of the whole?	6	
Can the children order amounts of money?	6	
Can the child tell number stories and identify the operations used?	7	